Bering Sea

36
35

34

37 **39**
38

40

Arabian
Sea
24

Bay of
Bengal

25
26

29

Philippine
Sea

32 **33**

31

PACIFIC

OCEAN

27
28

INDIAN

OCEAN

Coral
Sea

30

Tasman
Sea

22 U-995	**27** Houston	**32** Ships lost in Bikini Atoll	**37** I-400, I-401, I-201, I-14
23 Edinburgh	**28** Perch	**33** Prinz Eugen	**38** Arizona
24 John Barry	**29** Flier	**34** Wahoo	**39** YOGN-42
25 Lagarto	**30** Ships lost off Guadalcanal	**35** Borneo Maru	**40** Ships lost off Midway Atoll
26 Prince of Wales & Repulse	**31** Truk (Chuuk) Lagoon	**36** Grunion	

HIDDEN WARSHIPS

For Tony and Kathleen

First published in 2015 by Zenith Press, an imprint of Quarto Publishing Group USA Inc., 400 First Avenue North, Suite 400, Minneapolis, MN 55401 USA

© 2015 Quarto Publishing Group USA Inc.
Text © 2015 Nicholas A. Veronico

The information in this book is true and complete to the best of our knowledge. All recommendations are made without any guarantee on the part of the author or Publisher, who also disclaims any liability incurred in connection with the use of this data or specific details.

We recognize, further, that some words, model names, and designations mentioned herein are the property of the trademark holder. We use them for identification purposes only. This is not an official publication.

Zenith Press titles are also available at discounts in bulk quantity for industrial or sales-promotional use. For details write to Special Sales Manager at Quarto Publishing Group USA Inc., 400 First Avenue North, Suite 400, Minneapolis, MN 55401 USA.

To find out more about our books, visit us online at www.zenithpress.com.

ISBN: 978-0-7603-4756-0

Library of Congress Cataloging-in-Publication Data

Veronico, Nick, 1961-
  Hidden warships : finding World War II's abandoned, sunk, and preserved warships / Nicholas A. Veronico.
    pages cm
  Includes bibliographical references and index.
  ISBN 978-0-7603-4756-0 (hardcover with jacket)
  1. World War, 1939-1945--Naval operations. 2. Warships--History--20th century. 3. Shipwrecks--History--20th century. 4. Underwater archaeology. 5. Warships--Conservation and restoration. 6. Warships--Salvaging. I. Title.
  D770.V47 2015
  940.54'5--dc23
                    2015001525

Acquisitions Editor: Erik Gilg
Project Manager: Madeleine Vasaly
Art Director: James Kegley
Layout Designer: Helena Shimizu

On the front cover: High angle view of the bow of the *Rosalie Moller*, World War II shipwreck. Red Sea, Egypt. *Mark Doherty/Shutterstock*

On page 4: The USS *Arizona* Memorial. *Brett Seymour/NPS*

Printed in China

10 9 8 7 6 5 4 3 2 1

HIDDEN
WARSHIPS

FINDING WORLD WAR II'S
ABANDONED, SUNK,
AND PRESERVED WARSHIPS

NICHOLAS A. VERONICO

ZENITH
PRESS

CONTENTS

ACKNOWLEDGMENTS

Hidden Warships is a tribute to many dedicated people whose actions have added to the historical record of World War II. And although their individual motivations vary greatly, they are united by a common interest. For the divers, their motto is "Take only pixels, leave only bubbles." That is why, more than seventy years after the beginning of World War II, there are still many interesting objects to see on the sunken ships of the era.

Many surviving ships would not be floating without the cadre of trained volunteers of every craft who work so hard to maintain them. Their efforts to pass their trades down to today's up-and-coming shipwrights, electricians, mechanics, masters, sailors, and deck hands are an endeavor worthy of our financial support. The websites of many preserved ships are listed in the resources section, and those and other vessels deserve your support.

I would like to express my gratitude and thanks to the following people and institutions for their generosity in sharing their time and materials that made this work possible: the Abele family—Bruce, Susan, and John; Bob Alton; Bobby Anderson; David Anderson; Jerry Anderson; Hector Bado; Dan E. Bailey; Caroline and Ray Bingham; Wally Boerger; Joan Burke, Project Liberty Ship; Roger Cain; Tom Cates; Paul Chedotal, the U-Boat Story/Merseytravel; Ron Close; Tom Colgan; James P. Delgado; Jim Dunn; Jérôme Espla; Caroline Funk; John Geoghegan; Ellen Gerth, Odyssey Marine Explorations; Erik Gilg, Nichole Schiele, and Madeleine Vasaly, Zenith Press; Jerry Gilmartin; Laura Givens, Historic Ships Baltimore; Doug Grad; Kevin Grantham; Tiffany Gwynn, and John Royal, Naval Historical Center; Charles Hinman, USS *Bowfin* Submarine Museum and Park; Alexander Hub, Deutscher Marinebund e.V.; Terry Kerby, and Steve Price, Hawaii Undersea Research Laboratory; Daniel J. Lenihan; Taras Lyssenko; Joe Mazraani; Dale Messimer; Bob Mester; Randy Peffer; Stan Piet; Christophe Resse; Elizabeth Ruth-Abramian, Los Angeles Maritime Museum; Lee Scales; Christopher Scott, SS *Lane Victory*; Brett Seymour, National Park Service; Bradley Sheard; Dick Sleeter; Parks Stephenson; Patrice Strazzera; Ron Strong; Scott Thompson; Jeff Trentini; David Trojan; Rick Turner; Betty Veronico; Karen and Armand Veronico; John Voss; Thomas Weis, Wurttemberg State Library, Stuttgart, Germany; and Jennifer Wiggins, Auburn University Libraries—Special Collections/Archives. My sincere thanks. And, of course, any omissions or errors are my sole responsibility.

—Nicholas A. Veronico
San Carlos, California, 2015

INTRODUCTION

MUCH OF WORLD WAR II'S NAVAL HISTORY
LIVES ON AS HIDDEN WARSHIPS

In the years leading up to and during World War II, the Allied and Axis powers built more than 120,000 ships—ships of all sizes. During the conflict thousands of vessels were sunk, and after the war many more were scrapped or converted to perform new tasks.

The estimate of ships on the bottom of the world's oceans is staggering. US Navy combatants of all sizes, from water barges (classified as YW) to battleships (BB) to fleet aircraft carriers (CV), are counted at more than 1,875 vessels. The Allies sank 766 German submarines during the war, and Nazi U-boats in turn sent more than 2,775 Allied ships to the bottom. In the Pacific, the Allies sank 3,032 Japanese vessels, which displaced more than 10.5 million tons combined. The typical Japanese freighter, equivalent in size to an American Liberty ship, was 444 feet long with a beam of 58 feet and displaced seven thousand tons.

On the other end of the scale, America's largest combatant ships of World War II, *Iowa* class of battleships, displaced forty-five thousand tons and were 887 feet 3 inches long with a beam of 108 feet 2 inches. In contrast, Japan's super battleship *Yamato* displaced sixty-four thousand tons with an overall length of 862 feet 10 inches and a beam of 127 feet 7 inches. The Japanese lost all eight of their prewar battleships, and the two sister ships constructed during World War II, *Musashi* and *Yamato*, were sent to the bottom by Allied aircraft in 1944 and 1945, respectively.

In addition to the ship losses by the navies of the world, the United States put more than eighteen thousand vessels up for disposal as surplus after the war. These former combatants were given a new lease on life serving in a variety of civilian roles, from salvage tugs to transports to freighters moving goods for the new global economy that emerged after the war. Eventually, most were removed from service, many were scrapped, and some were hidden away, while others ended up on the bottom for one reason or another.

These vessels of war became today's hidden warships.

THE IMPACT OF TIME AND TECHNOLOGY ON WRECK DIVING

The ocean gives up its secrets only when it is ready. Yet the evolution of technology has enabled explorers to discover the ocean's secrets by hunting for, finding, and in some cases recovering a bounty of ships and treasure.

The digital revolution that began in the late 1970s changed sonar technology and camera miniaturization, and by the end of the twentieth century it made possible the use of remotely operated vehicles (ROVs) by sophisticated search and dive teams. In 2015, an advanced side-scan sonar can now be purchased for less than $10,000. When an interesting target is found, an equally advanced ROV, equipped with digital low-light still and video cameras, can be dropped over the side to investigate a contact. These technologies have greatly reduced the cost of shipwreck investigation and at the same time increased safety by eliminating the need to send divers down to investigate each and every contact. After investigating a target, how fast an ROV can be pulled up is only limited by the speed of the winch as opposed to the potential hazards of recovering a dive team, in which any number of factors can go wrong during ascent and decompression.

Beginning in 1969, Jacques Cousteau and his crew brought the realities of war into homes around the world through their explorations of Truk Lagoon and other former battlefields in the Pacific Ocean. Cousteau's 1971 TV special *Lagoon of Lost Ships* showed eerie scenes of gas masks and skulls inside sunken Japanese ships, contrasted with coral-encrusted cannon lit by the sun's rays beaming through crystal-clear water. Images of trucks and tanks on ship decks, holds with stacks of

Jarvis (DD-393) is one of the 32 Allied and 14 Japanese ships resting deep at the bottom of the Sealark Channel. She was sunk on August 9, 1942, with the loss of all 233 on board, the only US Navy surface warship to go down without any survivors. *Vallejo Naval and Historical Museum*

ammunition, and the crews' quarters with dinnerware scattered around the mess brought a wave of scuba diving tourists to Micronesia.

Two decades later, undersea explorer Robert Ballard brought a number of World War II sea battles back to life. Ballard discovered the wreck of the German battleship *Bismarck* in 1989. The Nazi battleship was remarkably preserved, having come to rest 15,700 feet below the Atlantic Ocean, approximately 475 miles west of the French port of Brest. An exploration of this magnitude had not been possible until side-scan sonar and ROV technology matured. At this point, however, both technologies were still extremely resource intensive in terms of cost, support staff, and equipment.

Turning his attention to the Pacific theater, Ballard in 1992 began focusing his research on the World War II naval engagements in the Solomon Islands. In addition to side-scan sonar and ROVs, Ballard's team also brought a three-person deep-diving submersible, *Sea Cliff*. The mini sub enabled close-up inspection of the wrecks and presented Ballard with the opportunity to take survivors of the battle down to see their former ships.

Ballard's team located the Australian heavy cruiser HMAS *Canberra*, which was pulverized in the opening minutes of the Battle of Savo Island (August 9, 1942). The heavy cruiser took twenty-four hits from Japanese guns, which immobilized it; after the battle, *Canberra* was abandoned and sunk by torpedoes from the destroyer USS *Ellet* (DD-398). As the battle raged, USS *Quincy* (CA-39) was mauled by Japanese gunfire and struck by a pair of torpedoes from the cruiser *Tenryū*. *Quincy* sank less than thirty minutes after the engagement began.

Ballard's expedition also located the American destroyers *Barton* (DD-599), *Laffey* (DD-459), and *Monssen* (DD-436), as well as the Japanese destroyer *Yūdachi*, which were all sunk during the First Naval Battle of Guadalcanal (November 1942). During that same engagement, the US battleship *Washington* (BB-56) hurled shells from its sixteen-inch guns into the Japanese battleship *Kirishima*, which capsized and sank on the morning of November 15, 1942. Also sent to the bottom on November 15 was the Japanese destroyer *Ayanami*. Ballard located this ship 2,700

A burning oil tanker sinks off the US East Coast, a victim of a German U-boat's torpedoes. During World War II, German submarines sank 2,775 ships. *Library of Congress*

Chicago (CA-29) is seen steaming in San Francisco Bay as she approaches Alcatraz Island. She had been damaged by a Japanese torpedo at the Battle of Savo Island on August 9, 1942, and was lost south of Guadalcanal and San Cristobal Islands on January 30, 1943, at the Battle of Rennell Island. *US Navy*

feet below the surface in two sections, apparently blown in two by an American torpedo that hit aft of the bridge.

Other ships resting in waters of the Solomon Islands include the American destroyers *De Haven* (DD-469), *Jarvis* (DD-393), and *Aaron Ward* (DD-483); cruisers *Chicago* (CA-29) and *Juneau* (CL-52); and the Japanese battleship *Hiei* and cruiser *Kinugasa*. In all, more than fifty ships were sent to the bottom during naval engagements in the area.

Fifty-six years after the June 4, 1942, sea battle around Midway Atoll, Ballard presented photos of USS *Yorktown* (CV-5) to the general public. The carrier is sitting nearly three miles below the ocean's surface with a 25° list to starboard. Both of the holes in *Yorktown*'s port side, made by torpedo hits from aircraft flying from the Japanese carrier *Hiryū*, were photographed by Ballard's team. Ballard's discoveries were turned into books and TV documentaries for the National Geographic Society, all consumed by a public extremely interested in World War II history.

Simultaneous to Ballard's expeditions in the Pacific Ocean, divers aboard Bill Nagle's boat *Seeker* discovered a submarine off the coast of New Jersey. Dropping down from *Seeker* were divers John Chatterton, Richie Kohler, and Kevin Brennan, who subsequently spent years trying to determine the German submarine's identity. Located 240 feet down, they repeatedly penetrated the wreck at great risk to themselves and eventually determined the sub's identity to be *U-869*. Many contested their findings, as it was thought that *U-869* had been lost off the coast of Africa. The U-boat's discovery, the subsequent attempts to determine its identity, and the deaths of three divers were profiled in Robert Kurson's best-selling book *Shadow Divers*. Chatterton and Kohler used their experience with *U-869* to develop the History Channel TV show *Deep Sea Detectives*, which ran for three seasons.

Wreck diving also preserved a president's reputation. During the 1992 presidential election, it was alleged that on July 25, 1944, young Ensign George H. W. Bush had attacked an unarmed ship and strafed Japanese seamen in a lifeboat. There was photographic evidence that Bush had dropped a five-hundred-pound bomb on the stern of an armed Japanese trawler, but for the political pundits, the photo and Bush's words were not enough.

As the George Bush "war criminal" story was about to break, divers Dan Bailey, Dave Buller, Pam and Chip Lambert, and Pat Scannon were en route to Palau, Micronesia.

After searching Kayangel Atoll, the reported scene of the alleged war crimes, the team quickly determined it was not deep enough to host ships, nor would it hold the wreck of a warship. Studying maps of local atolls, the team dived at nearby Ngeruangel Atoll, where they located Ensign Bush's target in seventy feet of water. Finding thousands of rounds of machine-gun ammunition as well as seventy-five-millimeter cannon shells on deck, the team videotaped the evidence that Bush had indeed attacked a Japanese combatant. Footage of the wreck put an end to the allegations.

From this trip, Pat Scannon went on to found the Bent Prop Project, a self-funded team dedicated to repatriating "every American service member who has not come home" from the World War II battles within the Palau Islands. Scannon and his group of volunteers have done tremendous work, discovering many of the missing from battles in this hotly contested archipelago.

These high-profile shipwreck explorations spanning more than three decades inspired the new millennium's generation of wreck-diving explorers. Each has given a tremendous boost to the general public's interest in World War II shipwrecks and wreck diving as a sport.

ON ETERNAL PATROL

During World War II, the United States lost fifty-two submarines, many of their locations unknown. In the past decade, half a dozen submarines that were once lost have been found, bringing closure to the crews' families and a final accounting for the missing submarines.

On May 18, 2005, USS *Lagarto* (SS-371) was found in the Gulf of Thailand by divers Jamie Macleod and Stewart Oehl operating from the dive boat *Trident*.

During May 3 and 4, 1945, *Lagarto* and USS *Baya* (SS-318) were tracking a Japanese convoy in the Gulf of Siam (today's Gulf of Thailand); convoy escorts harassed *Baya* to the point where the submarine broke contact, but *Lagarto* was never heard from again. The Japanese minelayer *Hatsutaka* had dropped depth charges on an unidentified submarine on May 4, with no results, and as it turns out, radar-equipped *Hatsutaka* had caught *Lagarto* and sent her to the bottom with all hands. The wreck lies 280 feet below the surface (at 7°55' N, 102°00' E), and divers have reported that one of her torpedo-tube doors is open and the tube empty. *Lagarto* did not go down without a fight.

The following year, 2006, saw the identities of a pair of once-missing submarines confirmed by the US Navy while a third submarine was

Video still from June 16, 2006, showing the plaque placed on the aft capstan of the World War II submarine *Lagarto* (SS-371), sunk in the Gulf of Thailand. Divers from USS *Salvor* (ARS-52) conducted six days of diving to positively identify *Lagarto* as part of the Thailand phase of the exercise Cooperation Afloat Readiness and Training, or CARAT. CARAT is an annual maritime training exercise involving the United States and six Southeast Asia nations designed to enhance the operational readiness of the participating forces. *Senior Chief Navy Diver Michael Moser/US Navy*

located in the Java Sea. In July, Russian divers confirmed a sea-bottom anomaly to be the missing *Gato*-class submarine USS *Wahoo* (SS-238), which was skippered by Cmdr. Dudley Walker "Mush" Morton. During his four patrols onboard *Wahoo*, Morton and crew sank nineteen enemy ships totaling more than fifty-four thousand tons.

Wahoo's final resting place and the grave of its eighty-man crew were located in the La Pérouse Strait between Japan's Hokkaido and Russia's Sakhalin Island, 213 feet below the surface. Finding *Wahoo* was an international collaboration led by the Wahoo Project Group—headed by Bryan MacKinnon, Morton's grand-nephew—the Sakhalin Energy Investment Corporation, Russian authorities, the Japanese Maritime Self-Defense Force, and the USS *Bowfin* Museum in Pearl Harbor, which consulted on the project. When *Wahoo* was discovered, the Russians were actually looking for their lost submarine *L-19*, which is believed to have been sunk by Japanese mines in the same area. On October 31, 2006, the US Navy confirmed that the submarine located in La Pérouse Strait was in fact *Wahoo*.

The forward capstan of the submarine *Lagarto* (SS 371), one of four missing US Navy submarines located and positively identified in 2006 and 2007. *Chief Diver Jon Sommers/US Navy*

In August, a location effort headed by the sons of Lt. Cmdr. Mannert L. "Jim" Abele, commanding officer of USS *Grunion,* proved fruitful. *Grunion* had been lost in early August 1942 near Kiska in the Aleutian Island chain. Photos from the expedition's ROV answered the questions as to why the submarine went down and verified its identity. (See the chapter on USS *Grunion,* page 154.)

At the end of 2006, on November 23, USS *Perch* (SS-176) was discovered in waters near Surabaya, Java. The submarine was found using sonar aboard the

Wahoo (SS-238) departs Mare Island Naval Shipyard, Vallejo, California, on July 14, 1943, prior to the World War II submarine's sixth patrol. On October 11, 1943, nearly a month into *Wahoo*'s seventh patrol, a multi-hour combined sea and air attack involving depth charges and aerial bombs sunk the *Gato*-class submarine. *Naval Historical Center*

Underwater view of *Wahoo*'s deck gun. The submarine was located 213 feet below the surface of La Pérouse Strait in July 2006. *Vladimir Kartashev/USS Bowfin Submarine Museum and Park*

dive boat *Empress*, owned by diver Vidar Skoglie. Kevin Denlay, Dieter Kops, Mike Gadd, and Craig Challen dived with Skoglie to the wreck. They located a plaque on the sub's hull confirming its identity. *Perch* was found in 190 feet of water, and it had been scuttled by its crew on March 3, 1942, after being heavily damaged by Japanese depth charges. The entire crew of five officers and fifty-four sailors survived the sinking, only to be interned for the duration of the war in Japanese prisoner-of-war camps. Six men subsequently died in Japanese captivity.

Three years after the spate of submarine discoveries, USS *Flier* (SS-250) was located in spring 2009 in the Balabac Strait, which separates Balabac Island from the northern tip of Borneo. Departing for her second war patrol from Fremantle, Australia, *Flier* made contact with an underwater mine on August 13, 1944. The submarine went down fast with only fourteen of the crew escaping. Of those fourteen, six perished during the long swim to a nearby island.

Flier was found by the crew of YAP Films, which was filming the documentary show *Dive Detectives*, hosted by father-and-son divers Mike and Warren Fletcher. When the submarine's identity was confirmed by the US Naval History and Heritage Command, Warren Fletcher said, "The *Flier* discovery presented *Dive Detectives* with one of our most challenging dives. At a depth of three hundred and thirty feet there is little margin for error. As my father and I descended into the dark blue water, the unmistakable shape of a *Gato*-class submarine came into view. That moment made all of the hard work and danger pale in comparison with the feeling of pride it gave me to know that *Flier* and her crew will not be forgotten."

On June 12, 1943, the training submarine USS *R-12* (SS-89) was lost in waters off Key West, Florida. The forward battery compartment of the 186-foot-long submarine began leaking, and the leak could not be stopped; as flooding accelerated, the order was given to blow the main ballast tanks to keep the submarine afloat. Unfortunately, the effort failed and the weight of the water took the boat down, forever entombing forty of its crew and two Brazilian Navy officers. Two officers and three sailors were swept overboard when the submarine went down, surviving to be rescued from the sea hours later.

Project leader Tim Taylor, Christine Dennison, and the crew aboard the research vessel *Tiburon* obtained a permit to search for the submarine from the US Naval

SS *Gairsoppa* left Calcutta, India, in December 1940, with a load of silver bound for the United Kingdom. In early February 1941, the steamer was torpedoed by *U-101* as she approached Galway, Ireland. Eighty-four men perished in the attack, with the lone survivor spending almost two weeks in a lifeboat before being rescued. *Library of Contemporary History, Stuttgart*

History and Heritage Command and began looking for *R-12* in fall 2010. On October 10, 2010, the submarine was located in six hundred feet of water. In 2012, Taylor and *Tiburon* crew photographed and mapped the vessel's condition. Another World War II submarine's final resting place was logged, and another lost crew would not be forgotten.

WORLD WAR II'S TREASURE SHIPS

The first of the large-haul treasure ships from World War II was the recovery of $105.2 million* (£65.9 million) in gold bars from the bomb room of the British cruiser HMS *Edinburgh*. At the time of *Edinburgh*'s sinking, the gold was being transferred to the United Kingdom to pay for shipments to keep the Russian populace supplied with food and its military with ammunition.

The cruiser had been escorting convoy QP-11, sailing from Murmansk, Russia, bound for the United Kingdom on April 28, 1942. Two days later, *Kapitänleutnant* Max-Martin Teichert, commander of *U-456*, fired a torpedo that struck *Edinburgh* on the starboard side. The ship listed immediately. Damage-control crews were able to isolate the flooding, but while they tried to return *Edinburgh* to an even keel, *U-456* sent another torpedo that literally blew off the cruiser's stern.

Edinburgh was taken in tow by destroyers HMS *Foresight* and HMS *Forester*, which pulled the convoy's stricken flagship back to Murmansk. Escorted by

* 2014 dollars

Odyssey Marine Exploration was awarded the contract to salvage the wreck of *Gairsoppa*.
This view of the ship's deck is from the underwater archaeological survey used to
determine the wreck's identity. *Odyssey Marine Exploration*

four minesweepers, the slowly moving tow was constantly harassed by Nazi
torpedo bombers. Nearing Bear Island on May 2, *Edinburgh* was fired upon by
three German destroyers. *Foresight* and *Forester* cast off their tow to pursue the
attackers. *Edinburgh*, its damaged steering forcing it to sail in circles, scored hits
on destroyer *Hermann Schoemann* (Z7), damaging it to the point that its crew

An ROV manipulator arm lifts a bar of silver from a stack inside *Gairsoppa*. The recovery effort spanned multiple seasons and the Odyssey Marine Exploration team recovered more than 100 tons of silver; the salvors were awarded 80 percent of the recovered treasure. *Odyssey Marine Exploration*

scuttled her. *Foresight* and *Forester* were successful at driving off the two remaining destroyers, but not before a torpedo was fired at them. The torpedo missed, but it continued into the side of *Edinburgh* exactly opposite its first torpedo hole. The cruiser, now fatally wounded, had its crew removed by minesweepers HMS *Gossamer* and HMS *Harrier*. After gunfire from the minesweepers failed to send *Edinburgh* to the bottom, a torpedo from *Foresight* sank her.

Nine years after the end of World War II, in 1954, the British government put the cargo onboard *Edinburgh* out for salvage, but deteriorating relations between the British and the Soviet Union during the Cold War canceled that salvage effort. Nearly three decades later, in the early 1980s, diver Keith Jessop put together a consortium to recover *Edinburgh*'s gold. The ship was found in the Barents Sea more than eight hundred feet below the surface. On September 16, 1981, diver John Rossier loaded a twenty-eight-pound ingot, the first of 431 gold bars to be recovered, into a metal basket that was hauled to the surface. In 1986, salvage crews returned to the wreck and an additional twenty-nine bars were recovered. This brought the total to 460. Compared to the manifest, only five bars were unrecovered and unaccounted for.

On August 28, 1944, another treasure ship laden with silver was sent to the bottom. The SS *John Barry* was an EC2-S-C1 Liberty ship built at Kaiser's Portland, Oregon, shipyard and launched on November 23, 1941. Liberty ships were 441 feet 6 inches long with a beam of 57 feet and a draft of 27 feet 9 inches, displacing 14,245 tons. During the war, American shipyards built 2,710 Liberty ships. Laden with three million silver Saudi one-riyal coins minted in the United States, *John Barry* was steaming for Dhahran, Saudi Arabia, when she was torpedoed by *U-859* approximately 115 miles off the coast of Oman. The wreck settled approximately 8,500 feet below the surface, practically guaranteeing it a safe final resting place.

In 1989, a consortium headed by Sheikh Ahmed Farid al Aulaqi, Brian Shoemaker, and Jay Fiondella was granted salvage rights to the former American steamship. Together, the men formed the John Barry Group to recover the freighter's cargo. In October 1994, the recovery effort began, with more than seventeen tons of silver being brought to the surface.

The most recent World War II treasure ship to give up her bounty of precious metals is the 412-foot-long steamship SS *Gairsoppa*. In December 1940, *Gairsoppa* departed Calcutta, India, bound for the United Kingdom. Having crossed the Indian Ocean, *Gairsoppa* rounded the Cape of Good Hope and traveled north to Freetown, Sierra Leone. There she joined convoy SL-64 for the journey to Liverpool, England, on January 31, 1941.

Low on fuel and traveling in worsening weather conditions, *Gairsoppa* left the convoy to stop at Galway, Ireland, to replenish its stocks of coal before continuing on—without the protection of the convoy's escorts. Traveling alone, *Gairsoppa* crossed paths with *U-101*, which sent four torpedoes toward the steamer. One found its mark, hitting the ship in the No. 2 hold. Of the eighty-five men on board, only one survived, coming ashore after thirteen days in a lifeboat.

Sixty-nine years later, the United Kingdom Department for Transport put the recovery of *Gairsoppa*'s cargo up for bid. Odyssey Marine Exploration Inc. of Tampa, Florida, won the salvage rights to recover *Gairsoppa*'s silver on an eighty-twenty split basis with Odyssey Marine taking all of the risk but, if successful, claiming 80 percent of the recovered silver.

After extensive research and a number of exploratory dives, Odyssey Marine found SS *Gairsoppa* more than three hundred miles off the coast of Ireland and more than 14,100 feet below the surface. Recovery operations began on May 31, 2012, with 1,218 silver bars—equaling approximately forty-eight tons—recovered during that season. The following year, the Odyssey Marine team recovered another 1,574 bars, weighing more than sixty-one tons. In total, the team brought up nearly 110 tons of silver, which is believed to be 99 percent of the amount listed on *Gairsoppa*'s manifest. Having monetized most of the silver recovered, Odyssey Marine Exploration and the UK government are splitting nearly $80 million.

In a published interview, Odyssey Marine president Mark Gordon said his company has a list of more than one hundred ships that are known to have cargoes valued in excess of $50 million each sitting on the bottom of the sea. There's still a lot to be found, and as technology improves, more will be recovered.

Aside from the treasure value, in the coming years many of the shallow-depth historic World War II–era wrecks may become "adventure tourism" destinations as multi-seat passenger-carrying submarines become more popular.

THEY'RE STILL OUT THERE . . .

The sea continues to give up its World War II secrets, and if the research proves correct, another treasure ship of that era is on the verge of being recovered. Sub Sea Research, headed by Greg Brooks, is after what is reported to be one of the richest shipwrecks in the Northern Hemisphere, holding platinum and gold that once belonged to the Soviet Union. In 2008, Brooks discovered the wreck of SS *Port Nicholson*, a British freighter his team believes was carrying a pair of Soviet emissaries, 7.4 tons of platinum ingots, and 4,882 gold ingots weighing four hundred ounces each. The precious metals were being sent from the Soviet Union to the United States in payment for foodstuffs.

The 481-foot-long *Port Nicholson* was traveling in convoy XB-25 from Halifax, Nova Scotia, Canada, to New York on June 15, 1942, when she was struck amidships and in the stern by two torpedoes from *U-87* off Cape Cod, Massachusetts. The freighter *Cherokee* was also sunk during the same attack. Today, *Port Nicholson* rests on her starboard side in eight hundred feet of water. The ship's main deck is fouled with fishing nets, so salvors will have to enter the freighter through either the side or bottom to access its cargo of precious metals. Sub Sea Research has been recognized as the legal owner and salvor of the ship, and time and funding will write the final chapter detailing *Port Nicholson*'s history and its treasure.

During the darkest days of World War II, the Japanese sank the cruiser USS *Houston* (CA-30) on the night of February 28, 1942, during the Battle of Sunda Strait. Nicknamed "the Galloping Ghost of the Java Coast," *Houston* entered Banten Bay at the northwest end of Java along with the Australian cruiser *Perth*, as officers believed the Japanese fleet was headed away from them. The Allied cruisers came face to face with heavy cruisers *Mogami* and *Mikuma* while a squadron of Japanese destroyers steamed in behind *Houston* and *Perth* to seal off their escape route to the rear. *Perth* was targeted first, and gunfire and torpedo hits sank it within one hour.

As the lone survivor, *Houston* took her toll on the Japanese ships, heavily damaging three destroyers and sinking a minesweeper. The first of four Japanese torpedo hits reduced her headway, and the following torpedoes stopped the American cruiser completely. *Houston* soon capsized, taking more than half its 1,061 crew members with her. Only 368 men survived the sinking, and they spent the remainder of the conflict in prisoner-of-war camps. The story of *Houston*'s heroic standoff with the Japanese was not fully known until after the war, when the ship's crew came home.

Silver bar recovered from *Gairsoppa*'s hold, all cleaned up after more than sixty years at the bottom of the Atlantic Ocean. In addition to the bar's serial number, weight, and purity, at the top is the mark of His Majesty's Mint— Bombay. *Odyssey Marine Exploration*

In summer 2014, the US Navy and Indonesian Navy divers performed a joint exercise to survey the wreck of *Houston*, which by international standards is considered a war grave. The site is a popular wreck-diving destination, and US and Indonesian officials are working together to safeguard the site.

At least once a month, news reports and the Internet bring stories of sunken ships into homes and to mobile devices the world over. These shipwrecks span all eras, but if one is patient, word of the next World War II–era hidden warship will appear.

US Navy cruiser *Houston* (CA-30) steams in the San Diego Bay in October 1935. The cruiser was sunk on February 28, 1942, at the Battle of Sunda Strait at Java. *Houston* did not go down without a fight, damaging three Japanese destroyers and sinking a minesweeper before being sunk herself. *US Navy*

Kristen Bauer, deputy chief of mission at the US embassy Jakarta, Indonesia (top left); Marine Lt. Col. Miguel Avila; and Capt. Richard Stacpoole pass a wreath to sailors assigned to Mobile Diving Salvage Unit (MDSU) 1 during a commemoration ceremony for *Houston* (CA-30) aboard the Military Sealift Command rescue and salvage ship USNS *Safeguard* (T-ARS 50). *Safeguard*, its embarked MDSU, and Indonesian navy divers conducted a diving exercise on the wreck of *Houston* as part of Cooperation Afloat Readiness and Training (CARAT) 2014. *Mass Communication Spc. 3rd Class Christian Senyk/US Navy*

"SWIMMING DOWN EITHER SIDE, PORT OR STARBOARD, YOU COME UPON MAJOR DAMAGE PRETTY QUICKLY. WHEN THE FORWARD MAGAZINE EXPLODED, IT EXPLODED WITH THE FORCE OF ONE MILLION POUNDS OF DYNAMITE, OR HALF A KILOTON. EVERYTHING WAS BLOWN OUT FROM UNDERNEATH THE NUMBER-ONE TURRET AND TWO OF THE DECKS COLLAPSED. THE SHIP'S FORWARD HULL WAS ALSO BLOWN OUT FOUR OR FIVE FEET ON EITHER SIDE IN THE EXPLOSION."

—DANIEL J. LENIHAN ON THE EXPLORATION OF USS *ARIZONA*

THE
OPENING
SALVOES
OF WAR

ACCOUNTING FOR PEARL HARBOR'S JAPANESE MIDGET SUBS

DURING THE NIGHT OF DECEMBER 6, 1941, seven to fifteen miles off the southern side of the island of Oahu, Territory of Hawaii, five midget submarines were launched from specially modified I-boat fleet submarines. The midget submarines were tasked with attacking the American fleet inside Pearl Harbor during the air raid planned for the next morning. During and after the air raid, the mother submarines were to sink any American ships that attempted to sortie from the harbor in pursuit of the Japanese carrier fleet. All of the midget submarines' two-man crews were prepared to give their lives in service to Japan's emperor.

The mother submarines sent to Pearl Harbor were *I-16* (commanded by Lt. Cmdr. Kaoryu Yamada), *I-18* (Cmdr. Kiyonori Ōtani), *I-20* (Lt. Cmdr. Takashi Yamada), *I-22* (Cmdr. Kiyotake Ageta), and *I-24* (Cmdr. Hiroshi Hanabusa). These fleet submarines were modern, most launched between 1940 and 1941, and were capable of long-range missions, able to cruise up to fourteen thousand nautical miles at six knots. On the surface, the submarines could make a maximum speed of 23.5 knots, or 8 knots submerged.

Each *Kō-hyōteki kō-gata*, or "Type A" midget sub, was referred to by the mother submarine's number and the term *tou*. The Pearl Harbor attack midget submarines were manned as follows: *I-16tou*, crewed by Lt. Cmdr. Masaji Yokoyama and Warrant Officer Sadamu Uyeda; *I-18tou*, Lt. Cmdr. Shigemi Furuno and Sub-Lt. Shigenori Yokoyama; *I-20tou*, Lt. Akira Hirō and Warrant Officer Yoshio Katayama; *I-22tou*, Cmdr. Naoji Iwasa and Sub-Lt. Naokichi Sasaki; and *I-24tou*, Lt. Cmdr. Kazuo Sakamaki and Warrant Officer Kiyoshi Inagaki. The mother subs were 358.5 feet in length and had a 30-foot beam, carrying the 78.5-foot-long midget submarines on deck aft of the conning tower. To launch a midget sub, the mother sub surfaced to enable the two-man crew to board, check systems, and prepare for the mission. A third midget-submarine crewman, essentially the *tou*'s crew chief, maintained the mini sub's systems, helped with the launch, and maintained radio communications after launch. Submerging again, the mother submarine would cruise to the departure location, launch its midget submarine, and then patrol while the mini sub conducted its mission. If successful, the two submarines would rendezvous and leave the area. If necessary, the midget submarine's crew would abandon their craft and escape onboard the mother sub.

The Japanese midget submarine crews posed for a group photo just prior to departing for Hawaiian waters and their mission to attack Pearl Harbor. From left, *I-20tou*'s crew of Akira Hirō (seated) and Yoshio Katayama (standing); *I-16tou*'s Masaharu Yokoyama (seated, second from left) and Sadamu Kamita (standing, second from left); *I-22tou*'s Naoji Iwasa (seated, center) and Naokichi Sasaki (standing, center); *I-18tou*'s Shigemi Furuno (seated, second from right) and Shigenori Yokoyama (standing, second from right); and *I-24tou*'s Kazuo Sakamaki (seated, right) and Kiyoshi Inagaki (standing, right). *Burl Burlingame Collection via Parks Stephenson*

TYPE A-CLASS MIDGET SUBMARINE

Length	78.5 feet
Beam	6 feet
Draft	10 feet
Displacement	46 tons
Propulsion	600-horsepower electric motors driving counter-rotating propellers
Speed	23 knots surfaced, 19 knots submerged
Crew	2

MIDGET SUBMARINES IN ACTION

In the quiet, early-morning hours of December 7, 1941—at 3:42 a.m., to be precise—Ensign R. C. McCloy aboard US Navy minesweeper *Condor* (AMc-14) spotted something that appeared out of place. Near the entrance buoys to Pearl Harbor, there was a periscope in an area where US submarines were prohibited from traveling submerged. Around 3:55 a.m., *Condor* reported its sighting to the destroyer *Ward* (DD-139) by signal light. *Ward* pursued the contact but after an hour and a half was unable to locate it. *Ward* requested the sighting information from *Condor* again and renewed the search.

By 6:00 a.m., *Ward* had broken off the search and resumed its regular patrol pattern off the entrance to Pearl Harbor. Lieutenant William W. Outerbridge, skipper of *Ward*, understood the significance of a Japanese submarine lurking off the harbor's entrance, but he did not report the phantom sighting to higher command at Pearl Harbor's Fourteenth Naval District.

Thirty minutes later, the general stores issue ship *Antares* (AKS-3) arrived off Pearl Harbor with a barge in tow. Lying to near the harbor entrance awaiting a pilot, an object resembling a small submarine was sighted 1,500 yards off *Antares*'s

Ward (DD-139) was a World War I–era, 1,247-ton *Wickes*-class destroyer built at the Mare Island Navy Yard in Vallejo, California. The "four-piper," so named for its four funnels, was launched on June 1, 1918, and is seen here moored in the San Diego, California, harbor on August 14, 1920. *Naval Historical Center*

MIDGET SUB CREWS

Commander Crew

I-24tou	Lt. Cmdr. Kazuo Sakamaki	W.O. Kiyoshi Inagaki
I-22tou	Cmdr. Naoji Iwasa	Sub. Lt. Naokichi Sasaki
I-20tou	Lt. Akira Hirō	W.O. Yoshio Katayama
I-18tou	Lt. Cmdr. Shigemi Furuno	Sub. Lt. Shigenori Yokoyama
I-16tou	Lt. Cmdr. Masaji Yokoyama	W.O. Sadamu Uyeda

Lt. Cmdr. = Lt. Commander   W.O. = Warrant Officer

The gun crew from *Ward* credited with firing the first shot on the morning of December 7, 1941, stands by the 4-inch/50-caliber gun amidships on the destroyer's starboard side. From left, Boatswain's Mate 2nd Class R. H. Knapp; Seamen 1st Class C. W. Fenton, R. B. Nolde, A. A. De Demagall, D.W. Gruening, and H. P. Flanagan; Gunner's Mate 3rd Class E. J. Bakret; and Coxswain K. C. J. Lasch. *Naval Historical Center*

starboard side. *Ward* was once again called in to investigate, and at 6:33 a.m. a Navy PBY Catalina patrol bomber from squadron VP-14 marked the sub's location with smoke bombs. The destroyer ran down the track and began shelling the midget submarine (now believed to be *I-20tou*).

Ward's No. 1 and No. 3 cannon opened fire on the midget submarine. The former's shot was high, sailing over the conning tower of the Japanese sub. The shot from the No. 3's four-inch/50-caliber (4"/50) gun, manned by Naval Reservists from Saint Paul, Minnesota, was fired from 560 yards and hit the sub above the waterline near the hull and conning tower joint.

Mortally wounded by cannon fire, the submarine passed under *Ward*, which then dropped a pattern of depth charges that sealed the fate of the Japanese midget sub and its two-man crew.

At 6:51 a.m., Lieutenant Outerbridge reported to the Fourteenth Naval District: "We have dropped depth charges upon subs operating the defensive sea area." Apparently feeling this was not direct enough, he transmitted again two minutes later: "We have attacked, fired upon, and dropped depth charges upon submarine operating in defensive sea area."

Destroyer *Ward* had drawn the first blood of World War II, more than one hour before the air raid on Pearl Harbor began, but the significance of the morning's actions would not be known until later in the day.

As the attack raged in the skies above Pearl Harbor, *I-22tou* came under fire at 8:37 a.m. from *Curtiss* and *Tangier*. Both ships engaged the midget sub in the channel on the northwestern side of Ford Island. *Monaghan* (DD-354) ran down the contact, and as the sub flooded its tanks to submerge, the destroyer attempted to ram it. Two depth charges were dropped by the destroyer, and the sub was never heard from again.

Twenty minutes after the air raid ended, the light cruiser *St. Louis* (CL-49) sortied from the harbor. Near the entrance, her crew spotted two torpedo wakes. *St. Louis* changed course, and both torpedoes exploded on a reef near shore. At about that moment, the midget sub, now lighter without the weight of the two

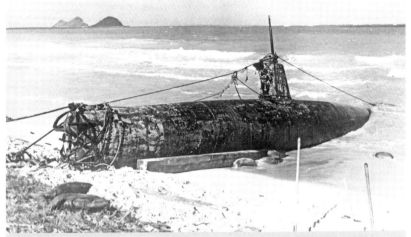

Sakamaki's *I-24tou* on the beach at Bellows Field the day after the attack. The midget submarine became stuck in the reef off the beach, but aerial attacks on the sub freed it from the coral, enabling it to float toward shore. It was pulled out of the water and secured using cables. In the following days, the sub was hauled out of the water, dismantled, and trucked to the naval base at Pearl Harbor for a complete inspection. *US Navy*

PEARL HARBOR MIDGET SUBS

Mother Sub	Mini Sub	Fate	Sinking Location	Discovery	Disposition
I-24	I-24 tou	Ran aground	Bellows Field, Oahu	8 Dec 41	National Museum of the Pacific War, Fredericksburg, Texas
I-22	I-22 tou	USS Monaghan (DD-354)	Pearl Harbor	21 Dec 41	Buried in pier, Pearl Harbor, Hawaii
I-20	I-20 tou	USS Ward (DD-139)	1,200 feet down, 3-4 miles off Pearl Harbor	28 Aug 02	In situ
I-18	I-18 tou	Depth charge, attacker unknown	Keehi Lagoon, Honolulu	13 Jun 60	Imperial Japanese Naval Academy, Etajima, Hiroshima
I-16	I-16 tou	Scuttled	Naval Defense Sea Area, outside Pearl Harbor	2000	In situ, discovered in 2000, confirmed in 2009

torpedoes, broke the surface. The cruiser's gunners opened fire and reported striking the sub, which they claimed sank immediately.

I-24's midget sub, commanded by Ensign Kazuo Sakamaki, began to have gyro-compass troubles and was unable to enter Pearl Harbor—no matter how many times Sakamaki tried. He eventually ran down the batteries' charge and was stopped dead in the water. He drifted most of December 7 and eventually ran aground on the opposite side of the island, near Bellows Field, the following morning. Though his crewmen drowned, Sakamaki swam to shore and was captured, becoming America's first prisoner of war of World War II.

ACCOUNTING FOR THE MIDGET SUBMARINES

In the years following the end of the conflict, historians tried to track the disposition of each of the midget submarines involved in the Pearl Harbor attack. The actions of some subs were known, while the sea held the secrets of others. The fourth and fifth submarines eluded historians for decades, and the final resting places of others raised more questions than they answered.

I-24tou was disassembled and studied at Pearl Harbor, examiners writing an in-depth report of the sub, its components, and its capabilities. It was subsequently shipped stateside, mounted to a low-boy trailer, and hauled around the country on a war bond tour. Eventually, *I-24tou* arrived at the National Museum of the Pacific War in Fredericksburg, Texas, hometown of Fleet Adm. Chester W. Nimitz, and placed on display.

At Bellows Field, a stone cairn and bronze plaque commemorate the midget submarine's capture and the fate of its crew. *Karen B. Haack*

I-22tou, which was sunk by the destroyer *Monaghan* inside Pearl Harbor, was raised

Destroyer *Monaghan* (DD-354) rammed and sunk *I-22tou*. As the destroyer passed over the midget sub, it dropped a series of depth charges. The midget submarine was raised on December 21, 1941, and it was clearly evident the depth charges had hit their mark. *Vallejo Naval and Historical Museum*

The midget sub sunk by destroyer *Monaghan* became fill when the Pearl Harbor submarine base was expanded. Above the pit and to the rear is *I-24tou* undergoing inspection and reassembly. *US Navy*

on December 21, 1941. It was visually evident once the vessel was dockside that it had suffered greatly from *Monaghan*'s depth charges; the shock waves reverberating on the shallow harbor bottom had intensified damage done to the submarine's hull. The bodies of Cmdr. Naoji Iwasa and Sub-Lt. Naokichi Sasaki were kept inside the submarine. Days after being recovered, *I-22tou* was used as fill in the expansion of the submarine base inside Pearl Harbor. Today, the sub is most likely buried underneath a parking lot.

Fifteen years after the Japanese surrender, in June 1960, *I-18tou* was discovered in seventy-five feet of water in Honolulu's Keehi Lagoon—just east of the mouth of the entrance channel to Pearl Harbor. The sub was missing its torpedoes, its hatches were unlocked from the inside, and the remains of the crew were nowhere to be found. The midget submarine was raised that summer, returned to Japan, and restored. It's now displayed at the Imperial Japanese Naval Academy at Etajima Island in Hiroshima Bay.

With the crewmen missing from *I-18tou*, many have wondered if they abandoned the submarine, swam to shore, and either blended in with the local population during the post–air raid chaos or met accomplices waiting to assist them to a safe house. Were they then to rendezvous with a Japanese submarine to be returned to the home islands? It should be noted that the midget submarine crews wore the same uniforms as the Japanese aircrews—was this an attempt to keep the midget-submarine operation secret, even if one or more of the crews were captured?

With *I-24tou* aground at Bellows Field, *I-22tou* sunk by destroyer *Monaghan*, and *I-18tou* at the bottom of Keehi Lagoon, that left *I-16tou* and *I-20tou* still unaccounted for. The latter is the midget submarine claimed to have been sunk by *Ward*'s crew. The sub obviously was indeed sunk, since it was never heard from again, but there were many detractors who said that the shot from *Ward* would have been impossible to make—a claim bolstered by how small a target the approximately five-foot-long conning tower of a Japanese midget submarine would make. Unable to find *I-20tou*'s final resting place for decades, many of *Ward*'s crew members passed away before the wreck was found and before they were given due credit for an incredible shot.

More than sixty years after she was sent to the bottom, on August 28, 2002, the deep diving submersibles *Pisces IV* and *Pisces V*, operated by the Hawaii Undersea Research Laboratory (HURL) located at the University of Hawaii's School of Ocean and Earth Sciences and Technology on Oahu, found *Ward*'s midget submarine. *I-20tou* was resting at a depth of 1,200 feet, three to four miles off the mouth of Pearl Harbor. It was clearly evident that this midget submarine had fallen prey to *Ward*'s gunners due to the distinct hole at the center of the base of the conning tower—right where the veterans said it would be. A better shot could not have been made. Vindication.

During a wartime overhaul, the Navy removed *Ward*'s No. 3 cannon, credited with sinking *I-20tou*. In 1958, it was presented to the state of Minnesota in honor of the contributions of its Naval Reserve, whose members manned the gun and made that fateful shot. The cannon was placed on the grounds of the state capitol for all to see.

Ward's gun crew had claimed that they put a round right through the Japanese midget submarine's conning tower; a claim many doubted. After sixty-one years, on August 28, 2002, the research submarines *Pisces IV* and *Pisces V* from the University of Hawaii's School of Ocean and Earth Science and Technology's Hawaii Undersea Research Laboratory (HURL) located the midget submarine five miles off the mouth of Pearl Harbor at a depth of 1,200 feet. The shell's hole can be seen at the base of the conning tower. *HURL*

I-16TOU: FATE UNKNOWN

For sixty-eight years, the fate of *I-16tou* was unknown. Through the years, HURL's operations director and senior pilot, Terry Kerby, and his team hunted for the both *I-20tou* and the unaccounted-for *I-16tou*, the fifth Pearl Harbor midget submarine. Said Kerby:

> In 1992, we began using our pre-science-dive-season test dives in an attempt to locate the *Ward*'s midget submarine. This came about from a meeting I had in 1991 with Dan Lenihan, then director of the National Park Service's Submerged

Parks Stephenson and *Ward* crewman Will Lehner with one of HURL's *Pisces* research submarines. Lehner was a fireman, second class, on board *Ward* during the December 7, 1941, sinking of *I-20tou. Courtesy Parks Stephenson*

Cultural Resources Unit [today known as the Submerged Resources Unit]. None of the federal agencies, the Naval Historical Center, National Park Service, or National Oceanic and Atmospheric Administration's National Marine Sanctuaries program could afford to pay for submersibles to wander around in the dark looking for historic wreck sites.

We could only use our first shallow test dive for these searches, and we would only get one dive a year [at most]. On top of that, we would not be able to do it every year. I did the first dive into the defensive sea area to start our search efforts in 1992, and located the tail section of what turned out to be the fifth midget submarine. We could see that it had been disassembled, rigged, and dumped, and assumed it was a war prize that was brought back to Pearl Harbor from the island-hopping campaign.

Eight seasons later, in 2000, we found the midsection, followed by the bow section in 2001, and we finally located the *Ward*'s midget sub in 2002. After the discovery, NOAA applauded our practice in using test dives to try to support other NOAA science missions and that encouraged us to make it an ongoing effort.

During each diving season's explorations, HURL's Steve Price accumulated a list of what historical artifacts were on the bottom in the waters off of South Oahu and their precise locations. From this, Price built a database of artifacts, which in subsequent years has enabled a number of historic maritime discoveries.

After the discovery of *I-20tou*, the HURL team noticed it was a carbon copy of the three-piece midget sub. Steve Price researched Japanese midget-submarine configurations and discovered that the figure-eight torpedo guard arrangement was only used on the Pearl Harbor attack subs. Kerby elaborated:

With Price's research, we then confirmed that we had, indeed, found the fifth midget submarine. From looking at the three sections, one can see that the sub had blown apart aft of the conning tower and we knew from studying the Sydney

Harbor midget sub that it was most likely from a scuttling charge. It had later been disassembled, rigged for disposal, and dumped in the defensive sea area.

We felt confident enough that we had located the fifth midget sub that we went public with the discovery on December 5, 2006, when HURL chief biologist Chris Kelley gave a talk in Session 19 of the Pearl Harbor symposium "A Nation Remembers," where he presented the Kerby-Price theory on the discovery of the fifth and final midget submarine.

The HURL team had found *I-16tou* outside of Pearl Harbor, and the evidence lying on the sea floor around the sub's parts gave them a good idea where it had initially sunk. After having located all of its components, the team then returned to the task at hand: undersea science missions.

The discovery of *Ward*'s midget sub generated a number of documentaries, one of which one was a History Channel production with diver John Chatterton. After it was completed, Chatterton expressed interest in working with HURL on future discoveries, and the information Kerby and his team developed on the fifth midget sub was forwarded to him. Chatterton, in turn, provided the HURL information to the Lone Wolf Documentary Group, which was working on a midget-sub production with Parks Stephenson, a graduate of the US Naval Academy (class of 1979), former submariner, and flight officer in E-2C Hawkeye airborne early-warning aircraft. Using the HURL data, Stephenson and Lone Wolf produced a PBS *NOVA* program titled *Killer Subs in Pearl Harbor*, which aired in 2010. The documentary team validated HURL's discovery by adding one final confirmation to the story of the fifth midget sub.

Having secured funding from the production company, Stephenson and the documentary team left for Japan to interview surviving submarine veterans. As Stephenson explained, one of the veterans had an interesting story to tell:

> Each of the midget sub-crews was a three-man crew. Two stayed in the sub and the third was, basically, the guy who took care of the sub while it was transported to the attack area on the back of the mother sub. Then, he saw the sub off, and he was the one to receive battle reports from the subs after they completed their mission. This gentleman, who was the only survivor of this cruise, said that almost toward midnight [on] the night after the attack, twelve hours after the last attacking Japanese aircraft had departed Oahu, he had received the pre-arranged signal from *I-16tou* reading mission success.
>
> Each of those subs had a discrete frequency, so he could only have heard a broadcast from his sub. This crewman, waiting aboard the *I-16* mother sub, had been instructed to listen on a certain frequency, and it was something like ten forty-five or eleven p.m., the night of December 7, [that] he received the *Tora* call. The submariners used the same codes as the aviators; we know this because there was a chart recovered from *I-24tou* that beached itself on the far side of the island. This chart gave the mission codes, and *Tora* was mission success.

The documentary crew's request to film the midget sub's wreckage was added to a NOAA-funded dive series with *NOVA* and Japanese TV network NHK picking up the cost of extending the mission by two days to include a visit to the fifth midget submarine. For the expedition, Stephenson brought a survivor from *Ward* and retired Japanese admiral Ueda Kazuo (JMSDF), who had been a midget submariner late in the war. Admiral Ueda descended to the wreck in *Pisces IV* with Maximillian "Max" Cremer piloting the research submarine, and it was he who gave the final validation of the HURL discovery.

"Diving down onto the wreck itself, the admiral concurred with the assessment that the details of the wreck were consistent with a Pearl Harbor attack midget sub and only to a Pearl Harbor sub," said Stephenson. "He brought mementos from the families of the two crewmen and offered them up in the submersible at the site. Kerby, who was piloting *Pisces V,* was able to get a scoop of sand from underneath the wreck. That was presented to the admiral, who took it back to Japan and presented it to the families. The Japanese were the final validation that we needed for the wreck, but [even] before they gave their validation, everyone involved was

Collaboration between Terry Kerby and Parks Stephenson put together the final fate of the fifth Japanese midget submarine from the Pearl Harbor attack.

pretty certain that it was the missing sub based on the physical details of the wreck [in particular the figure-eight torpedo guard at the bow]."

MAKING THE DIVE

Stephenson described the exploration of *I-16tou*:

In HURL's *Pisces* submersibles, we went down to about fifteen hundred feet. That was the depth of the Defensive Sea Area where the sub wreck was. The submersible takes its own atmosphere down with it, so you climb in, shut the hatch, and then descend. Carbon dioxide that is exhaled is scrubbed onboard and the atmosphere is no different [from] a submarine, really. It's just all a lot more cramped.

What we saw when we got down there was fascinating because the Defensive Sea Area outside of Pearl Harbor was a dumping ground for decades. You can see all kinds of stuff out there. Mostly war materiel, but you'll see cars and trucks, boats and planes, and ammunition. Live ammunition just dumped into the water. I don't know how many sixteen-inch shells we went over, just lying on the bottom. Rounds and rounds of ammunition.

As we got near the submarine wreck, what really caught my eye were these LVT amphibious track vehicle wrecks. They would later play into the theory on what happened to the midget sub and the circumstances under which it was found.

We got to the wreck of the midget submarine and confirmed that it was in three pieces, which was unusual. The midsection was blown apart. It looked like an internal explosion because the metal was bent outward and the location corresponded to the position of the scuttling charge that was known to be aboard the subs.

The midget subs were constructed in three sections. There are large seams before the conning tower and aft of the conning tower where the bow, center, and stern sections are bolted together. They had been separated, but more than just separated, the bolts had all been cut. Because we found evidence of bolts that were still concreted into their holes, it was evident that the Navy didn't just unbolt it and disassemble the sub. It had been underwater for a while and flooded, allowing the bolts inside holding the extensions together to become concreted in place.

Then, against that evidence were two empty torpedo tubes, which show no evidence of concretion, at least not with a torpedo in its tube. That would later lead us, after we had a forensic analyst who is [an] expert in shipwrecks and corrosion, assess that the torpedoes had left the tubes before the sub had been sunk.

The fact that the sub was broken up into three pieces and the salvage cable that was used to dump it over the side was still run through each of the pieces was another clue. Navy salvors actually ran the cable through the open end of one section and then punched a hole in the hull to help lift the sub's sections over the side.

The juxtaposition of the midget sub and the LVT-2s hinted that maybe the US Navy found the sub after the West Loch explosion in 1944. During the Pearl Harbor attack, there were some indicators of sub activity heading toward the West Loch area from eyewitnesses, and then finding these LVT-2s drew it all together. We have no documentation, of any kind, mentioning the recovery of this sub, the examination of this sub, or the disposition of the sub. The US Navy obviously raised it and disposed of it, but there's no documentation for that.

I-16TOU AND THE WEST LOCH DISASTER

Kept secret for more than fifteen years, the West Loch Disaster occurred in May 1944 and was not made public until 1960. LSTs (Landing Ship, Tank) were being loaded in Pearl Harbor's West Loch in preparation for Operation Forager, the invasion of the Mariana and Palau Islands.

In the May 21 incident, twenty-nine LSTs were moored in the West Loch and were being loaded with ammunition and supplies for the 2nd and 4th Marine Divisions' amphibious assault of the islands. At 3:08 p.m., *LST-353* exploded and started a chain reaction, causing nearby ships to add to the conflagration. *LST-39*, *LST-43*, *LST-69*, *LST-179*, and *LST-480* were sunk, and four additional LSTs were heavily damaged but later returned to service. More than fifteen LVTs and a number of 155-millimeter howitzers were also destroyed.

The final theory on the fifth midget submarine's final voyage is that *I-16tou* made a successful attack during the Pearl Harbor raid, then navigated itself into West Loch. With the harbor entrance sealed off by a now-vigilant American military, *I-16tou*'s crew sent the successful attack code and either abandoned their craft or activated the scuttling charge and perished in the resulting explosion. The midget submarine sank to the bottom of West Loch, only to be discovered by the US Navy when clearing debris—sunken LSTs and LVTs—from the area after the May 1944 disaster. The sub was hauled aboard a barge, sectioned into three pieces at the production breaks, then hauled out to sea and dumped with the remains of the West Loch incident. What remains unanswered is what happened to the midget submarine's crew.

MIDGET SUBMARINES IN SITU

Today, *I-20tou* and the components of *I-16tou* are jointly managed by NOAA's National Marine Sanctuary Program, HURL, and the National Park Service with assistance from the Naval History and Heritage Command.

Having the lead role in managing the two midget subs, NOAA has funded dives to the sites using the *Pisces* submersibles, the most recent in 2013. After sitting in corrosive saltwater for more than seventy years, both subs are beginning to deteriorate, and changes in their condition are being meticulously documented. The torpedo guard on *Ward*'s midget sub has deteriorated and fallen into the sand while the aft section is separating at the production joint, and the tail has separated and fallen to the sea floor.

The fifth midget sub is losing its battle with the elements as well. The hull is failing, and the conning tower will soon topple from the hull. Other small artifacts have been cleared from the conning tower's shadow and moved to safer locations so they will not be eventually crushed. For example, the aft running light that had deteriorated and fallen to the sand was picked up, placed in a bucket and tagged with the artifact's identification, filled with clean sand, and moved to a safer location on the wreck.

As the months go by, NOAA's Marine Sanctuaries program and the other stewards of the Pearl Harbor midget subs are developing plans for how to manage the wrecks, increasing the public's awareness of them while respecting that one or both are war graves.

Five months after the midget submarine attack on Pearl Harbor, the city of Sydney, Australia, was targeted by the two-man submersibles. During the night of May 31–June 1, 1942, a trio of Japanese midget subs entered Sydney Harbor. Two were detected and attacked; a third sent torpedoes after the cruiser *Chicago* but instead sank the Australian depot ship *Kuttabul*. The midget submarine seen here was recovered from the waters off Guadalcanal and is today displayed at the Submarine Force Museum in Groton, Connecticut. *Library of Congress*

REFLECTING UPON THE JAPANESE SUBMARINE THREAT

US naval forces in the Hawaiian Islands were constantly training to meet an expected threat: that of a submarine attack. It was theorized at the time that Japanese submarines would lie in wait at the mouth of Pearl Harbor and torpedo a number of ships in the harbor's entrance channel, thus bottling up the fleet. With the US Navy unable to sortie from the harbor, the Japanese would then invade Oahu.

Taking Oahu would have given the Japanese a staging base for an attack on the military and industrial targets on the West Coast at San Diego, Los Angeles, and San Francisco, California; Portland, Oregon; and Seattle, Washington. The Japanese could also have used Pearl Harbor to block shipping to Australia and other Allied nations, and a full-scale invasion of the US West Coast could have become a possibility as well.

"If one studies the Pearl Harbor battle, it was total success for antisubmarine warfare forces, and a total failure for the Japanese submarine force," said Parks Stephenson. "The first shot of the attack was by the destroyer *Ward*, and its target was a submarine. So the first shot, and the first victory, of Pearl Harbor went to the United States during the battle. After the attack, the American destroyers sortied, and they aggressively kept down and neutralized the Japanese fleet submarines that were all positioned outside the harbor. We were ready, but we weren't ready for the right threats, and that's what caught us off guard."

SNEAK ATTACK

THE DESTRUCTION
OF USS *ARIZONA*

EARLY ON THE MORNING OF SUNDAY, DECEMBER 7, 1941,
US Army privates Joseph Lockard and George Elliott were nearing the end of their duty shift at the Opana Point aircraft warning station on the northern point of Oahu. General Walter Short had put the island's Army radar stations, and the information centers they report to, on alert during what he considered "the most dangerous hour of the day for an air attack, from 4 o'clock to 7 o'clock a.m. daily."

Brett Seymour films the three fourteen-inch gun muzzles of turret No. 2 on *Arizona. Naomi Blinick/NPS*

Although slated to return to base for breakfast at 7:00 a.m., Lockard kept the radar set on to give Elliott some additional training while the pair waited for their ride back to base. Moments before 7:00, Lockard saw a huge formation on the radar's screen. Thinking the machine had developed a fault, he readjusted the set and then determined that it was working properly. At 7:02 a.m., Lockard began tracking a large formation of aircraft 132 miles north of Oahu, heading toward the island from 3° east.

While Lockard worked the radar screen, the information center shut down promptly at 7:00 a.m., and its crew headed for breakfast. Only Pvt. Joseph McDonald, a telephone operator, and Lt. Kermit A. Taylor, a fighter pilot sent to the center for liaison training and observation, remained on duty. Private McDonald received a call from Lockard at 7:20 a.m. and, with no one of authority on duty, transferred the call to Lieutenant Taylor. Hearing Lockard's report, Taylor told him "Don't worry about it," believing the radar was tracking a flight of twelve B-17s expected to arrive that morning from Hamilton Field, California.

"AIR RAID, PEARL HARBOR. THIS IS NO DRILL"

As it turned out, Privates Lockard and Elliott were tracking the first inbound wave of 189 aircraft that had been launched from Japanese aircraft carriers 260 miles north of Pearl Harbor at 26° north longitude, 158° west latitude. Surprisingly, Lockard and Elliott did not detect the pair of Mitsubishi A6M "Zero" floatplanes, one each from the cruisers *Tone* and *Chikuma*, sent in advance of the main strike force to reconnoiter Pearl Harbor and the Lahaina Roads anchorage on the northwestern coast of the island of Maui. *Chikuma*'s aircraft sent a report back to the strike force's carriers giving the number and types of aircraft stationed at Pearl Harbor as well as an early-morning weather report. *Tone*'s floatplane could only report that no American naval vessels were in the deep-water anchorage at Lahaina—where the Japanese preferred to attack rather than in the shallow waters of Pearl Harbor. Battleships *Hiei* and *Kirishima* as well as *Tone* and *Chikuma* each launched a Type 95 reconnaissance floatplane (Nakajima E8N) to patrol around the Japanese fleet during the attack.

Two waves of attacking carrier aircraft had been launched, the second following forty minutes behind the first. As the first wave crossed the Oahu coast, flight leader Lt. Cmdr. Mitsuo Fuchida signaled to his flight that they had achieved complete surprise. At 7:49 a.m., he gave the signal to begin the attack. Four minutes later, he announced to the fleet, "*Tora, tora, tora*," confirming the attack's surprise as the Zero fighters descended to strafe the air field at Ford Island and Nakajima B5N "Kate" torpedo bombers targeted the battleships and cruisers in the harbor.

ACHIEVING AIR SUPERIORITY

While the torpedo bombers roared in, low over the water, nine Aichi D3A "Val" dive-bombers focused their attention on aircraft parked in rows in front of Ford Island's Hangar 6. All airfields on Oahu had been alerted to the possibility of sabotage attacks by General Short, who had ordered all aircraft parked in rows, wingtip to wingtip, to make them easier to guard. Short also ordered that the bullets be removed from each aircraft's machine guns at nightfall.

Ford Island was home to Patrol Wing 2's PBY Catalina seaplane bombers and also served as the overhaul station for all Navy carrier-based aircraft. Thirty-three of the seventy aircraft on Ford Island were destroyed in the opening seconds of the attack by a flight of dive-bombers. Bombs destroyed Hangar 6 and heavily damaged Hangar 38. From his headquarters on Ford Island, Rear Adm. Patrick N. L. Bellinger sent a message to all Navy commands, stating: "Air Raid, Pearl Harbor. This Is No Drill." The message was sent at 7:58 a.m.

The worst of the attack was still to come.

Around 8:30 a.m., fifteen Douglas SBD dive-bombers from VS-6, the US Navy scout-bombing squadron aboard the carrier *Enterprise* (CV-6), approached Ford Island during the middle of the first Japanese attack wave. Thirteen of them landed safely at Ford Island; one was lost to the Japanese and one diverted to land on the island of Kauai.

Simultaneous to the attack on Ford Island, Zero fighters descended upon Kaneohe Naval Air Station, on the eastern side of Oahu and home to thirty-six Catalina patrol planes. Three of the base's aircraft were on patrol when the Japanese arrived overhead. The remaining patrol seaplanes were moored in the bay, parked on the tarmac, or under repair in the base's hangars. The first attack lasted nearly fifteen minutes, just long enough to set half the aircraft ablaze. When the Zeros departed, squadron personnel went into action, attempting to save the undamaged aircraft and put out those on fire. While base personnel were fighting the fires, a second squadron of Zeros roared over the field, strafing the men and aircraft and scoring a direct bomb hit on Hangar 3, which destroyed the four PBYs inside. When the Zeros left for the last time, only six lightly damaged PBYs remained to greet the three that returned from patrol.

The Schofield Barracks and adjacent Wheeler Army Air Field, located in the center of Oahu on Leilehua Plain, came under attack at 8:02 a.m. by twenty-five Val dive-bombers. Of the more than 150 planes at Wheeler Field that morning, nearly eighty were parked wingtip to wingtip in rows only twenty feet apart. Diving down from five thousand feet, the Vals bombed Wheeler's hangars and returned for low-level strafing of the ramp and barracks area. This attack, too, lasted fifteen minutes. Then the US crews began the task of fighting fires and attempting to arm planes for combat. During the lull in the attacks, four P-40s and a pair of P-36s took off from Wheeler Field to engage the enemy.

The day before the Japanese attack, the fighters of Lts. George S. Welch and Kenneth M. Taylor had been flown to Haleiwa, a dirt strip on the north coast of Oahu that went untouched by Japanese fighters. Welch and Taylor raced north from Waikiki to Haleiwa, got off the ground, and downed four aircraft each. In all, Army Air Corps pilots shot down twelve aircraft that morning.

At 9:00 a.m., seven Japanese planes returned for a few quick strafing passes of Wheeler Field en route back to the carriers. A total of eighty-three planes were destroyed or heavily damaged at Wheeler Field.

Wheeler's auxiliary strip, Bellows Field, located on the eastern side of Oahu and south of Kaneohe Bay, saw only one Zero during the first few minutes of the attack. This aircraft strafed the tent area and then flew off. At 9:00 a.m., nine more Zeros turned their attention to Bellows Field, destroying three of the twenty aircraft parked there.

Two airfields in the vicinity of Pearl Harbor were hard hit. Marine Corps Air Station Ewa Field, west of the entrance to Pearl Harbor, and the Army's Hickam Field, on the eastern side of the harbor, were rendered impotent within minutes

This captured Japanese photo of the opening moments of the attack on Pearl Harbor showing the hangars at Hickam Army Air Field on fire. In the foreground, torpedoes are headed for *Oklahoma*, and *Arizona* (second from left, foreground) is moored inboard of the repair ship *Vestal* (AR-4). *National Archives*

of the attack. Ewa Field was hit first by six Zeros that approached at one thousand feet and dived to within twenty-five feet of the ground to strafe planes and Marines attempting to defend the base. Since Ewa was on the way to Pearl Harbor, other Zeros and Vals made strafing passes at the field either en route to the target or when returning to the carriers. Marines broke out machine guns from the armory and were able to position one SBD dive-bomber for use as an antiaircraft gun mount. Ewa Field Marines are credited with downing one Zero during the battle. By the time the attack had ended at 10:00 a.m., nearly three-quarters of the forty-eight aircraft on Ewa's tarmac were ablaze.

Attacks on Hickam Field were well planned and precise. This airfield was home to the 18th Bombardment Squadron's long-range, four-engine B-17 Flying Fortresses and twin-engine B-18 Bolo bombers, which were perceived as a significant threat to the Japanese fleet. The first attack on the field lasted ten minutes and saw twelve Vals strike the Hawaiian Air Depot (a tenant of Hickam Field). As the machine shops and hangars of the depot exploded, seven more Vals strafed the flight line. At 8:25 a.m., another flight of Vals scored direct hits on the airfield's fuel pumping system, a number of the technical buildings, and the barracks. A third run on the field was made at 9:00 a.m., when nine aircraft strafed the hangar line and shop area while an additional half dozen machine gunned the living quarters, parade ground, and post exchange.

Arriving during the melee, the anticipated flight of B-17s from the 38th and 88th Reconnaissance Squadrons was descending to land after the long flight from California. To increase the range of the Flying Fortress, gasoline was loaded instead of ammunition for the bomber's defensive .50-caliber machine guns. Three of the 88th Squadron's bombers were brought into Hickam under fire, two landed at

A Japanese aerial photo showing the moment a torpedo exploded in the side of battleship *Oklahoma* on the far side of Ford Island. Moored on the near side, from left, are *Detroit* and *Raleigh* (both light cruisers), training ship *Utah*, and seaplane tender *Tangier*. *Utah* has already taken a torpedo hit and is beginning its list to port. *National Archives*

Haleiwa, and Lt. Frank P. Bostrom outran Japanese fighters and was finally able to set his aircraft down on a golf course. The 38th Squadron arrived between attack waves, and although one aircraft was damaged in the air, all were able to land safely. Of the twelve B-17s that arrived on the morning of December 7, one was destroyed (B-17C 40-2074), one was so heavily damaged it was salvaged for parts to keep others flying (B-17C 40-2049), and three were left damaged but repairable.

On this morning, the Japanese achieved total air superiority in the skies above Oahu. Less than two dozen American aircraft were able to sortie during the attack, with 188 destroyed and 151 heavily damaged. In comparison, the Japanese lost only twenty-nine aircraft and their crews.

COMBINED DIVE AND TORPEDO BOMBER ATTACKS

Signalmen on the ships at anchor in Pearl Harbor were hoisting the preparatory signal for the 8:00 a.m. raising of the colors when flights of dive-bombers could be seen approaching Ford Island. In the distance, torpedo planes were flying low as their crews set up attack runs. Twelve planes initiated the torpedo attack, flying into the harbor from the southeast, passing over the fuel tank farm, and heading directly for Battleship Row on the eastern side of Ford Island. Skimming the water at fifty feet, the Kate bombers dropped torpedoes fitted with wooden boxes around their fins to prevent the missiles from diving deep into the harbor and becoming stuck in the mud. The battleships moored on the outside row—*Arizona* (BB-39), *California*

A clock recovered from *Arizona* shows the estimated time the ship's forward magazine exploded, forever stopping time for 1,177 sailors and Marines lost on the battleship. This clock is displayed at the USS *Arizona* Memorial's Visitors Center. *Dave Trojan*

(BB-44), *Nevada* (BB-36), *Oklahoma* (BB-37), and *West Virginia* (BB-48)—were the intended targets on the first pass. *Oklahoma* took three torpedoes on the first pass and two more on the second; she began a severe list to port. As the Kates passed overhead, gunners in the rear cockpit strafed sailors on the ships.

The third torpedo attack came from the west, over Ford Island, and was launched at the minelayer *Oglala* (CM-4), berthed outside the light cruiser *Helena* (CL-50) at 1010 Dock (so named because it is 1,010 feet long). *Helena* and *Oglala* were berthed at the pier usually reserved for *Pennsylvania* (BB-38), which on this day was in dry dock. The single torpedo fired at the pair passed under *Oglala* and struck *Helena*, flooding one engine and boiler room and shorting the wiring to the main and five-inch batteries. Generator power was immediately restored to the turrets, which began engaging the low-flying aircraft. Captain R. H. English's crew aboard *Helena* isolated the flooding and was able to keep the cruiser afloat.

Concussion from the torpedo that impacted *Helena*'s hull split open *Oglala*'s hull plates and the ship began to take on water. Minutes later, a bomb was dropped between the two ships, knocking out power to *Oglala*'s pumps. The minelayer's crew abandoned her but enlisted the aid of a tug and moved the ship to the pier behind *Helena*. Two hours after the attack, *Oglala* capsized while tied to 1010 Dock. In spite of the damage, both ships would return to fight another day.

The last torpedo attack of the first wave flew low over the middle loch near Pearl City, aiming at ships moored on the west side of Ford Island. In their sights were *Utah* (AG-16), a World War I–era battleship that had been modified into a target ship; the light cruisers *Raleigh* (CL-7) and *Detroit* (CL-8); and the seaplane tender *Tangier* (AV-8). Torpedoes missed both *Detroit* and *Tangier*, but the others were not so lucky. *Utah* was moored at F-9, a spot usually reserved for aircraft carriers, and was struck by a pair of torpedoes in rapid succession. By 8:12 a.m., *Utah* had rolled over and sunk, taking sixty-four officers and men with her.

Detroit's crew fought gallantly to keep their cruiser afloat. A single torpedo flooded the No. 2 fire room and the forward engine room. Quickly counterflooding the listing ship, the crew worked to add additional lines to the mooring floats to keep her on an even keel. A Kate from the second wave dropped an armor-piercing bomb directly on *Detroit*, but the heavy shell penetrated straight through

A sailor aboard the hospital ship *Solace* (AH-5), docked across the harbor to the northwest of *Arizona*, used his movie camera with rare color film to capture the instant the battleship's forward magazine exploded. The explosion was estimated to have the force of one million pounds of dynamite. *US Navy via National Archives*

the lightly armored ship and exploded on the harbor bottom. *Detroit* was damaged but remained afloat.

Also during the second wave, the unscathed *Tangier*'s guns scored direct hits on three aircraft, which were seen to crash. Except for *Utah*, the remaining three ships—*Raleigh*, *Detroit*, and *Tangier*—would return to battle. *Raleigh* would be present at Tokyo Bay on September 2, 1945, for the Japanese surrender.

While the first wave's torpedo planes skimmed the harbor, dive-bombers were descending on the fleet followed by bombers in level flight passing over Battleship Row. At 8:06 a.m., *Arizona* was struck by a converted 1,760-pound battleship shell, which penetrated the main deck, descending into the forward fourteen-inch magazine and causing the gunpowder stored there to ignite. The magazine subsequently exploded, destroying the bow section of the battleship. The detonation instantly killed most of the men on board, including Rear Adm. Isaac C. Kidd, commander of Battleship Division 1 and the first flag rank officer to die in World War II, as well as the ship's commander, Capt. Franklin Van Valkenburgh. Both were posthumously awarded the Medal of Honor. More than half of all the American casualties during the Japanese attacks on the island of Oahu on December 7, 1941, occurred on *Arizona*.

Another Val dropped a bomb near *Arizona*'s funnel. A third bomb exploded on the boat deck, and a fourth hit the No. 4 turret. Four more bombs struck the superstructure amidships.

Lieutenant Commander Samuel G. Fuqua was the surviving senior officer onboard *Arizona*. He directed the damage-control efforts and the removal of the wounded from the ship's decks. Fuqua also gave the order to abandon ship and was one of the last to leave. He, too, was awarded the Medal of Honor for his actions. *Arizona* settled upright on the bottom of the harbor, taking forty-seven officers and 1,056 enlisted men with her.

Repair ship *Vestal* (AR-4), moored alongside *Arizona*, was rocked when the battleship exploded, blowing Cmdr. Cassin Young, the ship's captain, over the side. *Vestal* then took two bomb hits while Young was swimming back to the ship. Once back on board, Young moved his ship away from the burning *Arizona*, and with the assistance of harbor tugs he beached *Vestal* on Aiea Shoal. For his actions, Young was also awarded the Medal of Honor.

Tennessee (BB-43) was moored ahead of *Arizona* and inside *West Virginia*. Oil from *Arizona* was floating on the surface of the harbor, on fire, and threatened both *Tennessee* and *West Virginia*. Pinned between Ford Island and *West Virginia* with the sky obscured by thick, black smoke, *Tennessee*'s gunners could not see to shoot at any of the attacking planes. While the crew attempted to fight the fire on the water, the rear of the ship caught fire. High-flying Japanese bombers scored two hits on *Tennessee*, one each on top of Turret No. 2 and Turret No. 3. The bombs they dropped were converted fourteen- or fifteen-inch armor-piercing shells. When the attack was over, *Tennessee* was afloat, but salvage crews had to dynamite the forward mooring quay to untrap the vessel. She rejoined the fleet on December 20.

West Virginia was mortally wounded in the first torpedo attack by three torpedoes that struck below the ship's armor belt and one that impacted the belt. Two more are thought to have entered through the first torpedoes' impact holes while the ship was listing to port more than 20°. The resulting explosions gutted many parts of the battleship's aft interior spaces; in addition, the rudder was blown off the ship and was later found on the bottom of the harbor. High-level bombers dropped two fifteen-inch shells on *West Virginia*, but neither exploded and both were later found inside the ship. Burning fuel oil from *Arizona* entered *West Virginia* and added to the conflagration. The fires became so intense that the ship was abandoned, with the crew moving to *Tennessee* to help fight fires there. After the attack, *West Virginia* was refloated and prepared for the voyage to Puget Sound Navy Yard, Bremerton, Washington, for overhaul. She returned to the fleet two and a half years later, on July 4, 1944.

Forty minutes after the attack, *Oklahoma* graphically represented the destructive power of naval aviation. With the ship struck by at least five and possibly as many as seven torpedoes in the opening salvos of the attack, sailors attempted to isolate the resultant flooding, but it was too late. The ship immediately listed to 30°. As more bombs fell, the weight of the water inside the battleship increased. The crew began to abandon ship toward the starboard side as *Oklahoma* began to turn. At 8:32 a.m.,

continued on page 49

USS *UTAH* MEMORIAL

Lying on the west side of Ford Island, figuratively in the shadow of the *Arizona* Memorial and the battleship *Missouri* (BB-63), is the wreck of USS *Utah* (AG-16). The pre–World War I battleship had been converted to a target/training ship and was in port the fateful morning of December 7. As the attack got underway, *Utah* was hit by a pair of aerial torpedoes. The holes in the port side of the ship began to flood interior spaces, and the crew was unable to counteract that flooding. *Utah*'s list worsened, and most of the crew abandoned ship.

On March 9, 1909, the battleship *Utah* was laid down at the New York Shipbuilding Company's yard in Camden, New Jersey. The ship was launched on December 23 of that year and commissioned on August 31, 1911. The battleship saw service at the tail end of World War I, served as a gunnery trainer throughout the 1930s, and arrived at her new home port of Pearl Harbor in September 1941. *US Navy*

In July 1931, *Utah* was converted from a battleship to a mobile-target and antiaircraft training ship to comply with the 1922 Washington Naval Treaty. She was re-designated AG-16 and began her new career on April 7, 1932. Note that her twelve-inch cannon have been removed and that the turrets remain, but there are no barrels protruding. *US Navy*

SPECIFICATIONS: USS *UTAH* (BB-31/AG-16)

Length	521 feet 8 inches
Beam	88 feet 3 inches
Draft	28 feet 3 inches
Displacement	23,033 tons
Powerplant	4-shaft, 12-boiler Parsons steam turbines
Top speed	21 knots
Crew	1,001 officers and sailors
Armament	10 12-inch/45-caliber guns
Builder	New York Shipbuilding Corp.
Launched	December 23, 1909
Commissioned	August 31, 1911
Sunk	December 7, 1941
Location	Pearl Harbor, Hawaii
Coordinates	21.368955, -157.962455

Utah was one of the first ships attacked by Japanese torpedo bombers on December 7, 1941, rolling over and sinking at her berth. In 1950, two modest plaques were erected on the shoreline and on the ship. In 1972, a larger memorial was dedicated to the ship and her crew. *Ron Elias/ Northrop Grumman*

At 8:12 a.m., she rolled to port, taking sixty-four men who could not escape with her.

In the days following the attack, the Pearl Harbor Navy Yard's priority was to right the battleship *Oklahoma*; thus, *Utah* had to wait her turn. Beginning in the spring of 1943, using the parbuckling method to roll the ship upright, a salvage effort similar to that successfully used to right the *Oklahoma* was employed. *Utah* was 521 feet long with a beam of 88 feet 3 inches and a draft of 28 feet 3 inches. As the winches were rotating *Utah*'s hull, the ship slipped in the mud, moving toward Ford Island. By that point, she had been righted to a 38° list to port, and further salvage attempts were abandoned. *Utah* was left in place.

Small plaque-type memorials were placed on the hulk and later another on the dock near the ship. It was not until 1972 that *Utah* was given a full memorial to honor her and her shipmates who perished in the attack.

Beginning on October 23, 2014, the National Park Service, along with R2Sonic and Autodesk Inc., began a digital survey of *Utah* to determine the current state of the shipwreck. Using data collected from this survey, the Park Service plans to give a complete answer to the question "How long will the wreck last?"

Access to the USS *Utah* Memorial is currently limited to those with military identification cards at the west side of Ford Island.

The sun sets over the above-water structure of the battleship *Utah* (AG 16) at the USS *Utah* Memorial at Joint Base Pearl Harbor–Hickam on December 6, 2012. Recent renovations were implemented at the memorial to allow easier access for the public. *Mass Communication Spc. 3rd Class Diana Quinlan/US Navy*

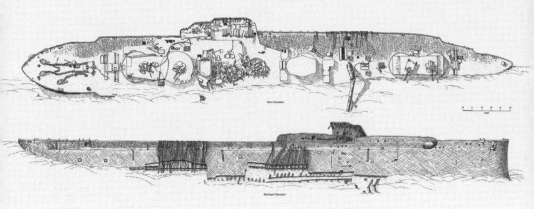

Starboard and port views of *Utah* (AG-16) after mapping by the National Park Service's Submerged Resources Center and the US Navy's Mobile Diving and Salvage Unit One (Detachment 319). The wreck does not look like much above the surface of the harbor, but there is still a lot to see and preserve on the hull. *Jerry Livingston/NPS*

her list to port became too great and the ship rolled over. In spite of heroic efforts to save men trapped within the battleship's hull, twenty officers and 395 enlisted men were lost when she overturned. Following the attack, *Oklahoma* was righted and salvaged, never to return to battle.

Berthed inside *Oklahoma* and pinned in the same predicament as *Tennessee*, *Maryland* (BB-46) was the most fortunate ship in the harbor. Because she was sheltered by *Oklahoma*, none of the torpedoes launched that morning struck her, and she was obscured by smoke from *Arizona*, *Tennessee*, and *West Virginia*. Most of the Japanese pilots went after more visible prey, and *Maryland* emerged from the attack virtually unscathed.

California was moored at F-3, directly across from 1010 Dock. The ship was slated for an inspection on Monday, December 8, and the manhole covers of six hatches into the ship's double bottom had been removed, the nuts of an additional twelve loosened. When two torpedoes struck the ship during the first wave, *California* quickly flooded. Ensign Edgar M. Fain ordered counterflooding that prevented the battleship from rolling over. Then a high-level Japanese bomber scored a hit amidships, which started a raging fire. As *California* slowly settled lower in the water, the floating fuel-oil fire from *Arizona* approached the ship. The fires became so intense that at 10:15 a.m., Capt. J. W. Bunkley, *California*'s commanding officer, ordered the ship abandoned. When the fires moved away, the order was rescinded. Three days later, *California* settled to the bottom of the harbor, listing 5.5° to port. *California* was raised on March 25, 1942, then dry-docked and repaired at Pearl Harbor. On June 7, 1942, she sailed east under her own power to Bremerton, Washington, for overhaul, rejoining the fleet on January 31, 1944.

Tanker *Neosho* (AO-23), moored at berth F-4, the gasoline wharf, had just finished unloading aviation fuel to tanks on Ford Island when the Japanese appeared overhead. As the fires spread around *California* and threatened the tanker, *Neosho* slipped her lines and backed away from Ford Island, seeking safer refuge and opening up an escape route for *Maryland*.

Tied up at berth F-8 behind *Arizona* and *Vestal*, battleship *Nevada* (BB-36) had steam in its boilers when the attack began. A torpedo hit the ship around 8:10 a.m. as bombers in the rear of the Japanese formation sought undamaged targets. A 4° to 5° list to port was addressed by counterflooding and making the ship watertight—known as "Condition Zed."

Burning fuel oil from *Arizona* began to engulf the water around *Nevada* when the senior officer present, Lt. Cmdr. J. F. Thomas, ordered the ship moved to a safer location. As *Nevada* steamed down the channel past *Arizona* and *Oklahoma* at approximately 9:00 a.m., five dive-bombers pounced on the battleship in an attempt to sink her and block the channel. One bomb hit the ship, passed through the side, and exploded in the harbor; two struck the forecastle; another exploded in the aviation gasoline tank; and the fifth bomb pierced the deck near the No. 1 turret.

Seven minutes later, more bombs rained down on the ship, one striking the port gun-director platform and another demolishing the crew's galley. Realizing that his ship may indeed block the channel if sunk, Thomas ran it into the mud off Pearl Harbor's Hospital Point. Tugs arrived and nursed the ship across the channel, where she sat with her stern aground and bow afloat. Four months after the attack, on April 22, 1942, *Nevada* departed Pearl Harbor for overhaul at Bremerton.

The second wave of attacks saw fifteen dive-bombers attack ships in the yard and dry-dock area, including the battleship *Pennsylvania* (BB-38) and the destroyers *Cassin* (DD-372) and *Downes* (DD-375). *Pennsylvania* was struck by a bomb that penetrated the main deck, amidships, starting a fire and taking the lives of two officers and sixteen enlisted men. Immediately, the dry dock was ordered flooded to within a foot of flotation in case the Japanese burst open the dock. Forward in the dry dock were *Cassin* (off the battleship's port bow) and *Downes* (off the starboard side). Around 8:50 a.m., ten to fifteen bombers approached the dock area. Destroyer *Shaw* (DD-373) in Floating Drydock No. 2 took a bomb hit to its forward magazine, which promptly exploded, blowing off the ship's bow. *Shaw* appeared to be a total loss; however, a patch was placed over her missing bow and the destroyer was sent to San Francisco, where a new bow was added. (She would return to Pearl Harbor in August 1942.) *Cassin* and *Downes* suffered a number of bomb hits that ignited magazines and stored torpedoes, and fires on one ship fed those on the other. *Cassin*

Arizona's no. 3 and no. 4 turrets were salvaged and installed at points on Oahu to be used as coastal defense artillery. Battery Arizona saw three of the ship's fourteen-inch guns and turret installed on the west side of the island at Kahe Point and Battery Pennsylvania, shown here, on Mokapu Peninsula in the hills above the Navy seaplane base at Kaneohe Bay (today known as Marine Corps Air Station Kaneohe Bay). The turret extended down more than seventy feet, and was supported by a magazine and crew spaces dug into the hillside. Note the turret ring gear at the top level of the barbette. *Dave Trojan*

SPECIFICATIONS: USS *ARIZONA* (BB-39)

Length	608 feet
Beam	97 feet 1 inch
Draft	33 feet 6 inch (max. draft)
Displacement	37,654 tons fully loaded
Powerplants	4 Parsons turbines, and 12 Babcock and Wilcox boilers
Horsepower	33,375 horsepower (max. speed of 21 knots)
Crew	92 officers, 1,639 enlisted (1,511 assigned on 7 Dec 1941)
Main Armament	12 14-inch/45-caliber guns mounted in 4 triple turrets (2 forward, 2 aft)*
Builder	Brooklyn Navy Yard
Launched	June 19, 1915
Commissioned	October 17, 1917
Sunk	December 7, 1941
Location	Pearl Harbor, Hawaii
Coordinates	21.364905, -157.949983
Website	www.nps.gov/valr/index.htm

* Each barrel was 52 feet 6 inches long and had a maximum elevation of 30 degrees. Each 1,400-pound armor-piercing shell was using four 105-pound powder bags. The muzzle velocity of the shells was 2,700 feet per second, and they were accurate to 34,000 yards (19.3 miles). Each gun had a rate of fire of 1.5 per minute.

rolled over to starboard, pinning *Downes* inside the dock. Both destroyers were subsequently returned to the fleet.

Seaplane tender *Curtiss* (AV-4) was another victim of a Val's dive-bombing attack. One bomb detonated on the main deck, taking the lives of twenty-one men and wounding another fifty-eight. Immediately exacting revenge upon the Japanese, *Curtiss*'s gunners scored direct hits on a Val that was pulling out of a dive. The tender's antiaircraft fire killed the pilot, and the plane flew out of control and crashed into *Curtiss*'s forward, starboard crane. The Val burned on the deck, destroying a large section of the ship's wiring, pipes, and steam lines. After repairs, *Curtiss* was returned to the fleet on May 28, 1942.

Just as the high-altitude bombers completed their mission at 9:15 a.m., twenty-seven Vals returned to strafe the harbor. Thirty minutes later, at 9:45 a.m., it was all over. Watching the Japanese aircraft fly north to rendezvous with the carriers, many stared at the utter destruction wrought during the one-hour, forty-five-minute attack. Although 2,280 military personnel were killed and 1,109 wounded, *Arizona*, *Oklahoma*, and *Utah* were the only vessels that would not return to the fleet. One hundred and eighty-eight hard-to-replace aircraft, however, were destroyed in the attack.

Despite these losses, the Japanese left much of Pearl Harbor's infrastructure intact. The fuel-tank storage farms were ignored, the submarine base went unscathed, and only three of the thirty destroyers present were knocked out of action. The Navy

An aerial view of the USS *Arizona* Memorial and the sunken ship below. Salvage efforts on the ship were suspended in the year after the attack when the ship was deemed broken beyond repair. *Arizona*'s bow is to the left and the sides of turret no. 2 can be seen above the surface of the water. To the right of the memorial are the barbettes for turrets no. 3 and no. 4. *Ron Elias/Northrop Grumman via NPS*

In 1983, Daniel J. Lenihan took the first-ever team of divers down to survey the *Arizona*'s hulk. At the time, Lenihan oversaw the underwater archaeology group, today known as the National Park Service's Submerged Resources Center. *Brett Seymour/NPS*

yard, hub of the rebuilding effort, also went undamaged, and—to the eventual death of many Japanese soldiers, sailors, and airmen—none of the US Navy's carriers were sunk.

WORKING ON AMERICA'S MOST HALLOWED SHIPWRECK

The Department of the Interior's National Park Service (NPS) is the lead agency in historic preservation within the US government. Well known for its stewardship of America's national parks, the agency manages 401 historic sites covering eighty-four million acres. These sites include 125 historical parks or sites, seventy-eight national monuments, fifty-nine national parks, twenty-five battlefields or military parks, eighteen preserves, eighteen recreation areas, ten national seashores, four parkways, four lakeshores, and two reserves.

In 1980, the USS *Arizona* Memorial and Visitors Center was transferred from US Navy control to the National Park Service. The agency's trained staff

Diver Deborah King maps artifacts on the deck of *Arizona*. From these recorded positions, detailed maps of the hull and an inventory of artifacts were made. Surveys conducted in subsequent years have been used to determine the rate of sediment accumulation on the wreck and to monitor the hull's deterioration. *Brett Seymour/NPS*

is able to interpret the battle and its remnants, including the wrecks of *Arizona* and *Utah*, for thousands of visitors each year. The memorial has been incorporated into the World War II Valor in the Pacific National Monument and receives approximately 1.4 million visitors each year.

When the memorial was turned over by the Navy, Park Service managers went looking for information. Everyone knew that when the memorial was built, the pilings were driven outside *Arizona*'s hull so that the structure is essentially a bridge over the shipwreck. Other than that, there were no maps and no detailed underwater photography, and no one who knew what the ship looked like in its present condition.

"Part of the reason for that lack of information, I think, is the harbor's low visibility," said Daniel J. Lenihan, retired National Park Service archaeologist and diver. "Typically it's about four or five feet. Sometimes it'll get up to six or seven feet, but the silt is very easily stirred up." Lenihan is the founder of the agency's underwater archaeology team known today as the Submerged Resources Center, which oversees and investigates a variety of underwater heritage sites. Beginning in 1983, he led the first-ever team of divers to map the *Arizona* wreck, a task that took several years.

"When you think about people looking at a map of a shipwreck, they're usually working on areas the size of a tennis court or a little larger—most likely a ship in the Mediterranean or something that might have gone down during the Civil War," Lenihan said. "*Arizona* is a site where if you started at the bow and swam around the ship, it would cover a distance of one-quarter of a mile. That is a big site."

To accurately map this big site, Lenihan and his team developed a technique that was new for the time but has since become standard practice. Using string and measuring tape, they ran baselines over the site and put markers down at important features in the wreckage. From any two points on the baseline, they could measure the distance to any marker. Those two distances, plus the distance between the two points on the baseline, formed a triangle. Called trilateration, this technique allows one to build a site map using simple geometry—it's like triangulation without angles. In murky water, it is much more accurate to measure distances than angles.

The trilateration system is self correcting; from the measurements, Lenihan's team could quickly tell if the points did not measure up. Another advantage of using this system is that additional divers, who were not archaeologists, could be employed, as their measurements could be verified without having to send another diver down to double-check the work.

National Park Service Submerged Resources Center chief Dave Conlin at the fourteen-inch guns of *Arizona*'s turret No. 2. Divers say that in the murky water of Pearl Harbor, objects tend to loom up out of the darkness. *Brett Seymour/NPS*

Preparing to dive on *Arizona* for the first time, Lenihan spoke to his team, relating that this ship was America's most revered modern war grave. As the nineteenth century has the Gettysburg battlefield, the twentieth century has Pearl Harbor, where World War II began for the United States, and USS *Arizona*, where—in the words of the National Park Service—the "greatest loss of life on any US warship in American history" occurred.

Said Lenihan:

We had the first uniformed archeological team in the Park Service, and part of it was because I anticipated us working on some sites [where] just for discipline and for appearance, we should all have a professional look. The entire *Arizona* dive operation was very disciplined and our people were very respectful. We had a very professional, older dive team—both Park Service and Navy divers—and everyone worked well together. The Park Service and the Navy take that site very serious, so we didn't have any trouble conveying that to our staff. The public could see that in the way we were treating the site, and we had a very good response from visitors when we were doing this work.

When you are out on the memorial or in the water around the ship, you are under a microscope considering every sound you make, anything you comment on. In the 1980s and 1990s, while we were working, it was possible to have guests on the memorial who might have been on this ship when it blew up, or were in the harbor when the attack occurred. Wounded veterans also come back to visit the memorial. Our staff had to show the utmost respect for the job and its location at all times.

The biggest problem the Park Service had was guests to the memorial were so fascinated with the divers and their work. There was concern that our presence would draw away from the reverence of the site and guests might feel we were an intrusion. But it seemed to go the other way—the biggest problem being they couldn't load guests back onto the boats because they wanted to talk to our staff at length. We had Park Service staff mapping in that little alcove area before you enter or depart the memorial, and they were attracting large crowds.

When the Park Service and Navy divers entered the water, they were diving on a ship with extremely limited visibility. It could get worse, but visibility on the wreck would rarely get better. When a person dives on a warship in clearer water, his or her mind is able to process what they're seeing and prepare to navigate through obstacles. On *Arizona*, however—with only five feet of visibility—the ship's structure has a tendency to "lunge out at you," according to Lenihan:

For example, when swimming on the forward part of the ship, the fourteen-inch muzzles just jump out, often startling you. One minute they are not there. You might be swimming forward and all of a sudden they're in front of you and you have to stop or swim around them. Swimming alongside of them is impressive. Coming to the end, the muzzles are fourteen inches across. If you're in the center looking at the gun tubes, you can see all three of them, but not much around [them], and it focuses your mind on them. It allows you to focus on details maybe more than you could in most situations. Being in low visibility, you've got no peripheral vision. The thing you're looking at you really see with all its complexity.

In diving on *Arizona*, one of the first things that struck me was the sheer size. It's a warship that's six hundred and eight feet in length and about one hundred and four feet in beam. Dropping down at the bow, I could actually wrap my hand around the cutwater. That ship came down to something that I could put my gloved hand around. Here I was at the bow, and the vessel's huge weight and all of its mass was before me. All of that mass is propelled across the ocean, slicing through the waves, yet here it rests on the bottom. It was very moving to float above the *Arizona* and have that revelation.

For the bow, swimming down either side, port or starboard, you come upon major damage pretty quickly. When the forward magazine ignited, it exploded with the force of one million pounds of dynamite, or half a kiloton. Everything was blown out from underneath the Number-One turret and two of the decks collapsed. The ship's forward hull was also blown out four or five feet on either side in the explosion.

continued on page 60

FLOATING PEARL HARBOR SURVIVORS

More than seventy years after the surprise attack on Pearl Harbor, it would seem a sure bet that none of the ships that were present or participated in the battle would still be afloat. But it would be a losing bet.

Three ships that were on duty that morning are still afloat—one still working, one in a museum, and one headed for a museum.

Still earning her keep is the former *YT-153*, now wearing the name *Hawk*. This 66-foot-2-inch-long tug was built at the Pearl Harbor Navy Yard and placed into service in 1941. Early on the morning of December 7, *YT-153* was sailing out through the Pearl Harbor Channel to deliver a pilot to the general stores ship *Antares* (AKS-3). As the attack got underway, *YT-153* was strafed by Japanese planes and soon after sighted a submarine periscope. The tug changed course to ram it, but the periscope submerged before the two vessels made contact.

YT-153 then rendered aid to the stricken ships in the harbor and helped beach *Nevada* and fight fires on board the battleship. At some point after the war, *YT-153* was shipped to the US East Coast and renamed *Hawk*. She now plies the waters of Narragansett Bay for Specialty Diving Services of Quonset Point, Rhode Island, moving construction barges from one point to another.

Moored in Baltimore Harbor for everyone to see is the white-with-orange-stripe Coast Guard cutter *Taney*. Through the years, the 327-foot-long ship saved many lives, and on the morning of December 7, she was berthed at her homeport of Honolulu Harbor just a few miles from Pearl. Her gun crews immediately engaged low-flying Japanese planes, and then the ship conducted anti-submarine patrols around Oahu following the attack. Later in her career, *Taney* escorted convoys in the Atlantic, fought in the amphibious battle for Okinawa in 1945, and in the late 1960s served in South Vietnamese waters providing gunfire support as well as medical aid to local civilians. The cutter was decommissioned on December 7, 1986, and is now permanently moored at the Baltimore Maritime Museum.

The last seaworthy Pearl Harbor attack survivor, *YT-146*

Docked at Pier Six, Honolulu Harbor, a few miles from Pearl, the Coast Guard cutter *Taney* was ready for action four minutes after the attack started. Using her three-inch guns, Taney opened fire on enemy aircraft flying over the area. The following day, *Taney* began antisubmarine patrols off the coast of Oahu. *US Coast Guard*

Hoga, recently emerged from dry dock and is preparing to move to its new home at the Arkansas Inland Maritime Museum. The *Woban*-class harbor tug was built at the Consolidated Shipbuilding Corporation in Morris Heights, New York, and commissioned on May 22, 1941. The tug was deployed to Pearl Harbor, and on the morning of December 7, 1941, she was underway not more than ten minutes after the first bombs fell.

It was *Hoga* that helped move the repair ship *Vestal*, moored next to *Arizona*, to safety. *Hoga* then moved the minelayer *Oglala* away from the light cruiser *Helena* (CL-50). Both ships were moored to 1010 Dock, and Oglala's hull plates had split from the concussion of a torpedo that passed under the minelayer and struck the cruiser; there was concern that if *Oglala* sank, she would pin *Helena* between the hulk and the dock. Shortly after *Hoga* assisted *Oglala* to a new mooring alongside 1010 Dock, the latter rolled over to port, sinking to the harbor bottom.

Although hard to see behind the bow of the battleship *Nevada* (BB-36), the tug *Hoga* (YT-146) is seen here pouring water onto the burning deck of the battleship. *Nevada* had escaped Battleship Row and was beached at Hospital Point to prevent her from blocking the harbor's entrance channel. During the attack and subsequent days after, *Hoga*'s crew worked non-stop putting out fires, moving ships, and rescuing sailors from the harbor. *US Navy*

Hoga is best known for fighting fires on the stricken battleships in the harbor; the photo on the previous page of *Hoga* spraying water onto the battleship *Nevada* beached at Hospital Point inside Pearl Harbor is iconic. The tug and her gallant crew went on to fight fires for

Hoga is the ultimate example of a historic ship hidden in the mothball fleets. After the war, she served as a fireboat for the city of Oakland, California, for fifty years. She is seen here at the far left, tied to a nest of ships at the Suisun Bay reserve fleet near Benicia, California. *Nicholas A. Veronico*

Riding the tides, *Hoga* sits in the reserve fleet lashed to the end of a row of ships on July 15, 2007. Who knew that this harbor tug (Navy designation YT) had played such a vital role in helping the Navy to recover from the December 7, 1941, attack on Pearl Harbor? *Nicholas A. Veronico*

seventy-two hours straight. After the war, in June 1948, *Hoga* became a fireboat for the city of Oakland, California, serving for more than fifty years. In December 1996, *YT-146* was placed into storage at the National Defense Reserve Fleet at Suisun Bay along the Sacramento River Delta at Benicia, California.

On July 25, 2005, *YT-146* was transferred on paper to the City of North Little Rock, Arkansas, to become part of the Arkansas Inland Maritime Museum. The tug was removed from the mothball fleet near San Francisco, dry-docked, overhauled, and painted, and it's currently docked at the former Mare Island Naval Base in Vallejo, California, awaiting a Navy barge that has the capacity to move it. *Hoga* is quite a historic ship for its size.

Restored and ready for display, *Hoga* waits for a tow tied up to the dock at the former Mare Island Naval Shipyard in Vallejo, California. The tug has recently emerged from dry dock and will soon be transported to the Arkansas Inland Maritime Museum in Little Rock, where she will be displayed in a place of honor. *Ron Close*

An open porthole reveals the division Marine office on the second deck of *Arizona*. *Brett Seymour/NPS*

Floating over the battleship's decks, Park Service divers found mostly utilitarian objects—the implements of daily shipboard life. Silverware, plates, bottles, and even a medicine cabinet that had vials and other items from the war were found on deck. Near the bow of the ship were a large number of loose .50-caliber rounds, possibly blown out of an ammunition locker during the attack. Each item found on deck was tagged, noted, and added to the wreckage maps. All of the measurements and drawings from the maps were used to develop the eight-foot-long model of the ship in its present condition that is now on display in the visitors' center.

In addition to mapping *Arizona*'s final resting place, Park Service divers were looking at the overhead spaces in the compartments to see where oil had collected. It should be noted that divers do not enter the ship's hull; they use remotely operated vehicles (ROVs)—cameras on poles extended into the interior spaces—so as not to violate the integrity of the

The flag section stateroom located on the second deck of *Arizona*, photographed through an open porthole. *Brett Seymour/NPS*

hull and to be respectful to the fallen sailors entombed there. Much of that effort is aimed at tracing how oil from shattered bunkers is moving through the ship, a major management concern. As the ship's interior evolves in its underwater environment, trapped oil migrates toward the surface. The Park Service monitors the

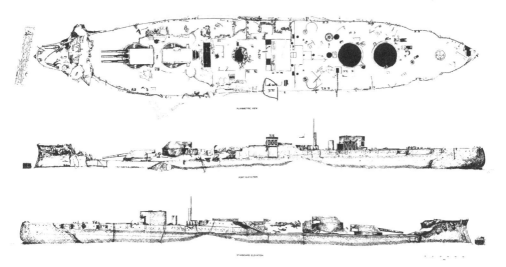

Above: National Park Service Submerged Resources Center and US Navy divers teamed to measure the *Arizona* wreck and catalog its features and associated artifacts. From the information gathered by the team, NPS artist and diver Jerry L. Livingston developed these renderings of the battleship. *Jerry L. Livingston/NPS*

Below: Submerged Resources Center deputy chief Brett Seymour films an open hatch on *Arizona's* main deck. Note the ladder inside the hatch. No divers have entered the *Arizona's* hull since salvage work concluded in 1943. *Naomi Blinick/NPS*

ship's oil outflow, reporting that somewhere between two and nine quarts of Bunker C oil escapes from *Arizona* each year. Many call the small oil droplets that rise from the wreck "black tears" or say that the ship is shedding the oil in remembrance of the crew still entombed within its hull.

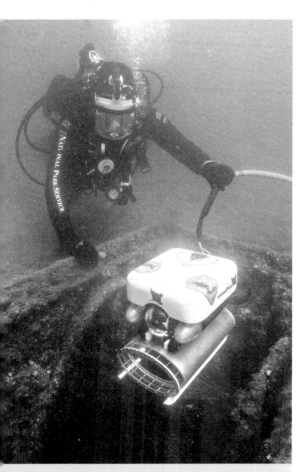

Dave Conlin deploys an ROV into an open hatch on *Arizona. NPS*

When the ship sank, there were more than 1.5 million gallons of Bunker C on board. However, the wreck burned for more than two days, consuming an unspecified amount of oil. "Many have asked why we haven't just drained the remaining oil to avoid a potential environmental disaster," said Lenihan. "The reason this has not been done to date is that there is not just one fuel tank, like in a car or motorboat, but more than a hundred bunkers spread through the ship, all in different stages of decomposition. To access them would require sending crews deep inside the ship, now considered a war grave. Also, our Navy dive partners at Pearl Harbor—Mobile Diving Salvage Unit One [MDSU-1], who are experts in salvage problems—tell us there would be little chance of success without making things a lot worse for quite a while." Battleship *Arizona* is a war memorial marking the beginning of World War II and serving as a symbol of the sacrifices made by men and women of the US armed forces in general and the ship's crew in particular. It is a war grave, being the site where 1,177 men gave their lives. And it also marks a time when the United States was prepared for a naval war yet unprepared to defend against a new weapon of war—airpower.

Today, National Park Service divers routinely inspect *Arizona*, doing so with the respect and reverence accorded a war grave and a war memorial.

Kelly Gleason, at the time an NPS volunteer diver, surfaces above *Arizona* surrounded by oil released from the battleship. Estimates place the amount of oil escaping from the ship at nine quarts per year. Gleason subsequently earned her doctorate and today is the chief scientist, maritime archaeologist/diver for the NOAA's Papahānaumokuākea Marine National Monument in the Hawaiian Islands. *Brett Seymour/NPS*

A moving image of the USS *Arizona* Memorial shot from the bow looking back to the center of the ship during an Oahu rainstorm. The water-level perspective is unique and could only be obtained by Navy or NPS divers. *Brett Seymour/NPS*

CAUGHT IN THE OPEN

HMS *REPULSE* AND HMS *PRINCE OF WALES*

IN AN ATTEMPT TO PREVENT THE JAPANESE from attacking British possessions in Borneo, Malaya, and the territory known as the Straits Settlements—which included Malacca, Penang, and Singapore, among others—the Royal Navy battleship *Prince of Wales* and battlecruiser *Repulse* were dispatched to Singapore as a projection of strength.

Known as "Force G" during their deployment from the United Kingdom, *Prince of Wales* and *Repulse* were accompanied by destroyers HMS *Electra*, *Encounter*, *Express*, and *Jupiter*, arriving in Singapore on December 2, 1941. The group was redesignated as "Force Z" upon arrival. The aircraft carrier *Indomitable* was to rendezvous with the Force Z ships at Singapore, but the carrier ran aground off Jamaica. She was in dry dock at the American naval base at Norfolk, Virginia, and did not make the passage to Malaysia. Thus, the Force Z ships had no sea-based air cover available to protect them.

In the early morning hours of December 8, 1941, Japanese high-altitude bombers attacked the British stronghold at Singapore. *Prince of Wales* and *Repulse* responded with antiaircraft fire but did not suffer any damage. Simultaneous to the bombing, twenty-four thousand Japanese troops landed at Kota Bharu, approximately 460 miles north of Singapore.

At 5:10 p.m., local time, Force Z (*Prince of Wales*, *Repulse*, and destroyers *Electra*, *Express*, *Tenedos*, and *Vampire*) sortied from Singapore Harbor to attack the Japanese invasion force. The following afternoon, December 9, the Japanese submarine *I-65* detected the British ships and trailed them, reporting their position as they moved north. Having received *I-65*'s reports, the invasion force's commander, Vice Adm. Jisaburō Ozawa, turned his now-empty transports back toward port at Cam Ranh Bay in Japanese-occupied French Indochina.

The British force was then located by scout planes from the cruisers *Kinu*, *Kumano*, and *Yura* at approximately 5:30 p.m. Shortly thereafter, destroyer *Tenedos*, being short on fuel, left Force Z and headed back to Singapore. Darkness fell over the British ships as they pursued the Japanese invasion force. A night attack was launched by Japanese land-based bombers, but they were unable to locate the British ships as stormy weather closed in.

At around 10:00 p.m., a patrolling Japanese seaplane mistook the heavy cruiser *Chōkai* for *Prince of Wales* and dropped a flare. As the Japanese naval force turned to the northeast, ships in Force Z spotted the flare as well. Admiral Sir Tom Phillips, commander of Force Z, immediately turned his ships to the southeast. Unable to locate the Japanese ships, Phillips decided to return to Singapore to await further developments.

At 3:40 a.m., Force Z steamed past *I-58*. The submarine gave chase, firing five torpedoes at the British ships, none of which found their targets. *I-58*'s position reports were picked up by the 22nd Air Flotilla, which prepared ten aircraft to search for the ships at first light.

During the night, Phillips received word that the Japanese had landed at Kuantan, about 250 miles north of Singapore; he turned his ships and headed in that direction. This had the unintended consequence of throwing off the Japanese search planes, which did not locate the ships of Force Z until 10:15 a.m. One hour later, eight twin-engine Mitsubishi "Nell" (G3M) bombers concentrated their attacks on *Repulse*. Seven 550-pound bombs missed, but one struck the battlecruiser in the starboard stern deck area.

During her short career, HMS *Prince of Wales* saw a lot of action. Built at Birkenhead, England, across the River Mersey from Liverpool, she was completed in March 1941. The ship was heavily damaged in the fight to sink the German battleship *Bismarck*, and after repairs went on to fight the Italians in the Mediterranean. *Prince of Wales* arrived in Singapore on December 2, 1941. Six days later, *Prince of Wales*, battlecruiser *Repulse,* and four destroyers encountered Japanese airpower in the South China Sea. *National Archives*

The battlecruiser HMS *Repulse* joined the British fleet in 1916, undergoing modifications in 1922 and again in 1933. The second modernization lasted three years and saw the ship fitted with improved deck armor and additional antiaircraft batteries. In May 1941, *Repulse* was part of the squadron that sank the *Bismarck*. In the fall of 1941, she was sent to protect British interests in the Far East. *Naval Historical Center*

Right: A Japanese photo from the opening attack on *Repulse*, bottom of frame, and *Prince of Wales*. Seven near misses can be seen in the waters around the stern of *Repulse*, and the dark area covering the starboard stern section is a direct hit. Neither ship had long to live. *Naval Historical Center*

At 11:45 a.m., two squadrons of Nell bombers fitted with torpedoes came in for the attack. Eight torpedoes were aimed at *Prince of Wales* and nine sped toward *Repulse*. All torpedoes, except for one, missed. One torpedo struck *Prince of Wales* at the junction where the outer propeller shaft exits the lower hull. This threw the propeller out of balance, chewing into the inner port propeller, which caused the prop shaft to break. The outer propeller fell away, and water rushed into the hull through the now out-of-round shaft opening. The in-rushing water caused the battleship to list 11.5° to port, which cut power to her pumps, the secondary armament, and steering. Powered by just her starboard engines, *Prince of Wales* was able to make fifteen knots, but directionally she was unstable.

While *Prince of Wales* was dealing with her problems, *Repulse* came under concentrated torpedo attack. Ten more Japanese torpedoes were aimed at the British battlecruiser, with one finding its mark on the port side. Two more torpedoes hit the ship as well. *Repulse* was built with a shallow antitorpedo bulge, and the explosion of the torpedo's warhead pierced directly into the side of the ship. Without the watertight compartmentalization of newer ships, *Repulse* was unable to stem the flooding. Within six minutes, the ship rolled over and sank.

Three more torpedoes found *Prince of Wales* and one 550-pound

A low-level photograph of Force Z's demise taken from a Japanese torpedo-bomber. At far left is *Prince of Wales*, with *Repulse* in the background, second from left. In the foreground is a British destroyer. *Prince of Wales* was heavily damaged in the first wave of torpedo attacks. *National Archives*

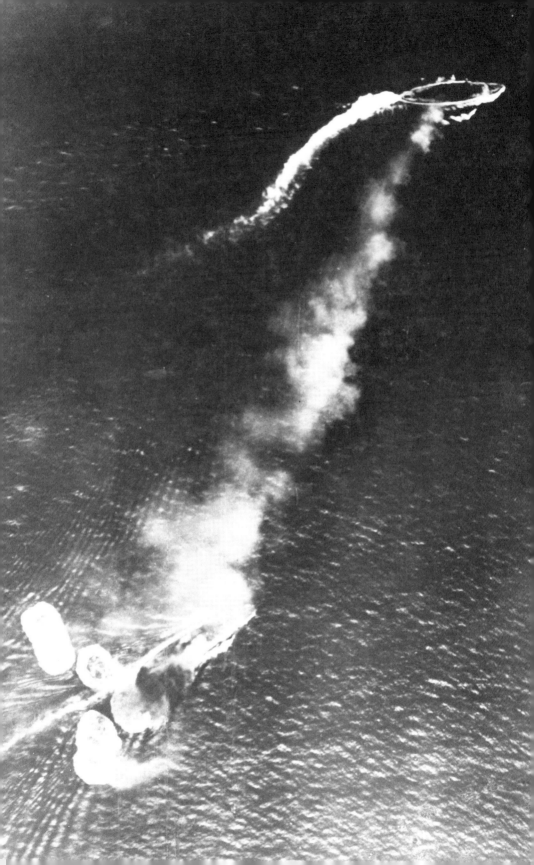

In this photo shot from the bridge of the HMS *Express*, the destroyer has put its hull next to *Prince of Wales* to enable the battleship's crew to jump from one ship to the other. *Prince of Wales* is listing heavily to port as men can be seen clambering over her side. *Author's collection*

SPECIFICATIONS: HMS *REPULSE*

Length (overall)	794 feet 1.5 inches
Beam	90 feet 1.75 inches
Draft	27 feet
Displacement	27,200 tons
Powerplant	2 steam turbines driven by 42 water boilers
Top speed	31.5 knots
Crew	968 officers and sailors
Armament	6 15-inch guns in 3 dual gun turrets 9 4-inch guns (triple mounts) 8 4-inch guns (dual mounts) 8 20-millimeter cannon 8 21-inch torpedo tubes
Class	*Renown*-class battlecruiser
Builder	John Brown & Co., Clydebank, Scotland
Launched	January 8, 1916
Commissioned	August 18, 1916
Sunk	December 10, 1941

SPECIFICATIONS: HMS *PRINCE OF WALES*

Length (overall)	745 feet 1 inch
Beam	103 feet 2 inches
Draft	34 feet 4 inches
Displacement	43,786 tons
Powerplants	Geared steam turbines driving 4 propellers
Horsepower	100,000 shaft
Top speed	28 knots
Crew	1,521 officers and sailors
Armament	10 14-inch guns 16 5.25-inch guns 32 40-millimeter cannon 80 antiaircraft rocket projectors
Class	*King George V*–class battleship
Builder	Cammell Laird and Co., Ltd., Birkenhead, England
Launched	May 3, 1939
Commissioned	January 19, 1941
Sunk	December 10, 1941

bomb struck the deck amidships, exploding below decks. As the battleship continued to list, destroyer *Express* pulled along her starboard side to aid the crew as they abandoned ship. At 1:18 p.m., *Prince of Wales* slid under the South China Sea. Both of the Royal Navy's largest capital ships in the region were gone.

DIVING THE *PRINCE OF WALES* AND *REPULSE* WRECKS

Today, the wrecks of *Prince of Wales* and *Repulse* lie eight miles apart, approximately 120 miles north of Singapore and fifty miles north of the island of Pulau Tioman, a popular resort diving destination. As a dive site, the scale of *Prince of Wales* is mind boggling. It's 750 feet long, and she was a quantum leap ahead of her contemporaries—more than two hundred feet longer than other World War I battleships.

Rod Macdonald, diver and author of *Force Z Shipwrecks of the South China Sea*, has made a number of descents to both *Prince of Wales* and *Repulse*. Macdonald's first trip to the Force Z shipwrecks was in 2001 at the invitation of the British military, serving as a consultant to the service's divers, who were testing the use of trimix gasses.

Diving the shallowest parts of *Prince of Wales* can be done on air, but at the ship's depth of approximately 225 feet, using trimix—a combination of oxygen, helium, and nitrogen—is safer and provides for more bottom time. The wreck of *Repulse* can be safely dived on air, as it is shallower, with the ship's starboard rail at 125 feet and the bottom at 180 feet. Although both wrecks can be dived the entire year, the best times are March and April and again in September through November, when the water is a comfortable 86° Fahrenheit and visibility can be more than 120 feet.

"On my first visit, we tied into *Prince of Wales* for a week of dives," said Macdonald. "At the wreck, there's a monsoon current that runs in one direction six months of the year, then it runs the other direction for the other six months. It's a very steady current at about two knots. We used underwater scooters that moved at two and a half knots, meaning that even in the current we could get around these wrecks quite easily."

Prince of Wales is lying upside down in 230 feet of water, and the bottom of the hull—the top of the wreck—comes to between 160 and 180 feet of depth. In addition, *Prince of Wales* sits beam-on to the prevailing current. The current sweeps along the seabed, hits this massive obstruction (the 750-foot-long hull), and stirs up all the sand and sediment. Said Macdonald:

> There's a lot of debris lying around the wreck on the seabed. It's a clean, white, sandy seabed, and artifacts lie where they fell when the ships sank. Because the current changes direction, there's not a great pile-up of silt against the hull covering it up—it's all swept away. It's very much the battle scene that it was when the ship went down. There's still a number of Carley floats [life rafts made from a copper tube covered with cork and canvas] recognizable on the bottom.
>
> The water on the way down to the wreck is very clear, with one hundred to two hundred feet of visibility. At the top of the wreck it's okay as well, but as soon as you go over the side to where the current is hitting the hull, it can be very cloudy and the visibility can drop down to ten to twenty feet. The sediment-filled water can have a milky sort of feel to it.

Most divers will descend to the top of the upturned hull, then drop over the side. It's a descent of another sixty feet along the sheer side of the battleship's hull to reach the rail along the main deck. Here, divers start to see all the recognizable features of the battleship—the superstructure and the secondary armament—albeit upside down. On most diving expeditions, the group heads to the most recognizable feature, the main batteries. "What is most impressive is their position," said Macdonald. "I was scootering along the side of the starboard ship looking for A turret, which is the one with the four big, fourteen-inch main guns. I ducked underneath the overhanging hull, and through the murk I could see a strip of light green one hundred feet in the distance. I realized when I looked around that the whole bow from the forward turret was actually suspended off the seabed by the strength of the armor and the turrets. The whole bow was one hundred to one hundred and fifty feet off the seabed and I was looking across the entire beam of the ship to the port side. It is an amazing sight to see."

Diving the wreck of *Repulse* is different. This ship is a battlecruiser, so it's the same length as *Prince of Wales* but not as big overall. The wreck is rolled more onto the port side, so it is not completely upside down like *Prince of Wales*. Her orientation on the bottom enables the monsoon current to run from bow to stern, and this

A diver maneuvers around the twin fifteen-inch gun barrels of *Repulse*'s B turret. The ship rests on her port side in 180 feet of water; her starboard side is only 125 feet below the surface, making this an easy descent for technical divers. Unfortunately, the wreck's shallow depth also makes *Repulse* and *Prince of Wales* a target for metal scrappers, who are, essentially, slicing and dicing the gravesites of more than eight hundred British sailors. The wreck lies approximately 120 miles northeast of Singapore. *Guy Wallis, courtesy Rod Macdonald*

sweeps all of the sediment away. "At *Repulse*, the underwater visibility is a staggering one hundred to two hundred feet of visibility," said Macdonald. "She's a far cleaner, silt-free wreck, and because she's more on her side you can see an awful lot more of her. The main guns are easy to see. The midship's conning tower and bridge super-structure and all the side armament are all recognizable and exposed. This wreck is more easily accessible as she's not quite as deep as *Prince of Wales*."

ILLEGAL SALVAGE ACTIVITY ON THE WRECKS

"There has been a considerable degree of illegal salvage activity going on over the last few years on these wrecks and the war wrecks of other nations in the area," said Macdonald.

During the battle, a torpedo struck *Prince of Wales* beside the bracket that holds the three-part propeller shaft that connects the hull to the propeller. Now out of balance, the shaft started vibrating and thrashing around inside the hull, eventually breaking the seventeen-inch shaft made of the best British steel. That propeller

The camera points down the twin fifteen-inch gun barrels of B turret out into the South China Sea. The range finder and rear section of A turret can be seen at right. Unfortunately, the large 15-inch guns were of no use in defending the ship from the high-level and torpedo-bombers of the Imperial Japanese Navy's 22nd Air Flotilla. *Guy Wallis, courtesy Rod Macdonald*

fell away to the seabed, and the battleship eventually sank with the three remaining propellers. Macdonald said:

> Each of these props is absolutely massive, probably fifty tons of phosphor bronze. They were lifted off the wreck [in 2013], and the two remaining props from the *Repulse* were also taken. Earlier this year, Malaysian authorities caught a salvage barge working Japanese war wrecks off the west coast of Malaysia and impounded it with four million dollars' worth of scrap metal on board.
>
> We thought that was the end of it, but the problem continues to grow. There was a party of divers at the wreck sites at the end of 2014, and they found a number of explosive charges on the hulls. Unscrupulous divers are setting charges to separate the hull plates so they can be removed and the steel recycled for money. Each of the hulls had explosives wired together with detonation cord in coffee tins strung together. And these things were all over the wreck and they were getting ready to fire.
>
> I think what's happened is that the salvage vessel that's working the wrecks had seen the dive boat approaching the site on radar and they left the scene, or had gone just over the horizon to hide. Once the dive vessel left after the dives were complete, as it was cruising away, the divers heard three loud explosions coming from the direction of the wrecks. The salvors had returned and fired the charges. The illegal salvage work is still ongoing; however, this salvage barge has now been seized. British divers are appalled by what has been happening because these are very sensitive British war graves. Hopefully these arrests will see an end to what has been happening—but a significant amount of damage has already been inflicted on these war graves.

To bring the problem closer to home, the damage being done to *Prince of Wales* and *Repulse* is equivalent to someone trying to salvage a hallowed American site like the battleship *Arizona* or strip-mine the Gettysburg battlefield. Do it and you'll have a fight on your hands.

Unfortunately, *Prince of Wales* and *Repulse* lie in international waters, which limits Britain's rights to protect the wreck sites. Both of the shipwrecks are considered "protected places" under the British Protection of Military Remains Act, but nationals of countries outside the United Kingdom do not have to honor the act.

"The Malaysian authorities are trying to stop this plundering, but the salvage barge comes from China or another nation that does not honor our war dead," said Macdonald. "It's somebody who we don't have positive relations with and all they see is metal that can be converted to money. They're actually salvaging Japanese and Australian warships as well. It's absolutely disgusting."

JUNE 4, 1943

11:22.5 A.M.: "U-BOAT SURFACED, BEARING 083°TRUE, 8,000 YARDS FROM THE CARRIER. TWO VFS [WILDCAT FIGHTERS] STRAFED AND ESCORTS OPENED FIRE."

11:23 A.M.: RADIO TRANSMISSION FROM COMMANDER, TASK GROUP 22.3—CAPT. DANIEL V. GALLERY—TO COMCORTDIV 4 [COMMANDER, ESCORT DIVISION FOUR]: "I WOULD LIKE TO CAPTURE THAT BASTARD IF POSSIBLE."

11:24 A.M.: ESCORT CARRIER *GUADALCANAL*: "LAUNCHED KILLER GROUP OF ONE VT [AVENGER TORPEDO BOMBER] AND 1 VF [WILDCAT FIGHTER]."

11:25 A.M.: DESTROYER ESCORT "*CHATELAIN* REPORTED U-BOAT HIT SEVERAL TIMES. *JENKS* ORDERED TO CEASE FIRE."

11:27 A.M.: "*CHATELAIN* REPORTED, 'THEY ARE ALL HOLDING THEIR HANDS UP. THEY ARE SURRENDERING.'"

—*WAR DIARY*, TASK GROUP 22.3, JUNE 4, 1944, ABOARD THE ESCORT CARRIER *GUADALCANAL* (CVE-60), GIVING THE MINUTE-BY-MINUTE ACCOUNT OF THE CAPTURE OF THE GERMAN *U-505*

№ 2

HUNTING
THE KILLERS

SALVAGE WORK ON THE *ADMIRAL GRAF SPEE*

WHEN HISTORY LOOKS BACK AT *KAPITÄN-ZUR-SEE* HANS Wilhelm Langsdorff, it sees a gallant man, a gentleman, and a man who cared about his crew. Langsdorff commanded the German navy's *Panzerschiff* (armored ship) *Admiral Graf Spee*—pronounced "grahf shpay"—during its only war cruise from August 23 to December 17, 1939.

Langsdorff steamed from Germany's North Sea port of Wilhelmshaven on August 23, traveling north along the coasts of Denmark and Norway, west toward Iceland, and then down to the South Atlantic to avoid the British Isles and any long-range patrol aircraft the British might have scouting the area. War in Europe began on September 1, 1939, and that same day the heavy cruiser met its replenishment ship, *Altmark*, north of the equator in the middle of the Atlantic Ocean. *Altmark* provided oil and spare parts, and the two ships cruised together heading south. During the ensuing two weeks, *Altmark* and *Graf Spee* conducted training drills and provided cover for each other while both ships, in turn, came to a complete stop to work on their machinery.

German Naval Warfare Command (*Seekriegsleitung*, or SKL) radioed Langsdorff on September 25 that Royal Navy cruisers *Ajax*, *Cumberland*, *Dispatch*, and *Exeter* and destroyers *Hotspur* and *Havock* were patrolling the coast of South America. The next day, SKL gave *Graf Spee* permission to begin offensive operations in the area. Three days later, on September 30, Langsdorff's first victim sailed over the horizon. *Graf Spee* confronted the 5,050-ton British steamer *Clement* sailing south of Recife, Brazil. The Booth Steamship Line's freighter hailed from Liverpool, England, and was surprised when *Graf Spee* steamed directly for her bow while transmitting orders to stop and not to send a distress call.

In spite of the instructions from the heavy cruiser bearing down on the ship, *Clement*'s radio operator sent a distress call with the ship's position. A boarding party from *Graf Spee* put *Clement*'s men into their lifeboats and pointed them to the nearest large port, Maceio, Brazil, which the crew reached the next day. *Clement*'s captain and chief engineer were held prisoner aboard *Graf Spee* and interrogated by the Germans. Langsdorff fired two torpedoes at *Clement*, but neither exploded (or they missed), and he opened up with cannon fire. After *Clement* went under, a Greek-flagged freighter, *Papalemos*, passed in the vicinity and was ordered to stop, which it did. Both of *Clement*'s officers were transferred to the Greek freighter, and the cruiser then departed the area. *Graf Spee* had drawn its first blood.

To put some distance between the sinking position of *Clement* and its next victim, Langsdorff sailed east toward the African continent. Steaming independently, the freighter *Newton Beech* was caught as she transited from Cape Town, South Africa, to Freetown, Sierra Leone (British West Africa), with a cargo of corn

Stern view of *Admiral Graf Spee* prior to her first war patrol in the spring of 1939. Note the large Nazi eagle beneath the rail at the stern. *Hector Bado collection*

destined for England. At Freetown, *Newton Beech* was to join a convoy, but she never made it. Langsdorff took the ship as a prize of war and held the crew prisoner. Along the same route on October 7, *Graf Spee* captured the freighter *Ashlea*, her holds full of sugar, also en route for a rendezvous with an England-bound convoy at Freetown. *Ashlea*'s crew was put aboard *Newton Beech*, and *Graf Spee*'s engineers blew holes in the newly captured freighter in order to sink her. The next day, October 8, Langsdorff took both crews from *Newton Beech* on board *Graf Spee*. *Newton Beech* was then sunk by explosive charge.

Graf Spee headed back toward the mid-Atlantic and in doing so crossed paths with the Charente Steamship Co. Ltd.'s 8,196-ton freighter *Huntsman* on the morning of October 10. Unable to send a raider report, the steamer was taken as a prize with her crew transferred to *Graf Spee* with the cruiser's boarding crew now sailing *Huntsman*. For eight days, *Huntsman* shadowed *Graf Spee*. On October 18, the two ships met up with *Altmark*, and cargoes and prisoners were exchanged. When the transfer was done, *Huntsman* was scuttled with explosives.

Trevanion was built in 1937 for the Hain Steamship Co. The relatively new diesel-powered freighter displaced 5,299 tons and was 432 feet long, with a beam of 56 feet 2 inches and a draft of 24 feet 8 inches. The ship was sailing from Port Pirie, South Australia, to Swansea, England, and she, too, was to join a convoy at Freetown for protection on the voyage north. She never joined the convoy. *Graf Spee* captured the freighter, took the crew prisoner, removed what stores they needed, and then scuttled the ship.

After sending *Trevanion* to the bottom, Langsdorff set sail for the Indian Ocean. En route, the German cruiser met *Altmark* at a point one hundred miles northeast of Tristan da Cunha, an archipelago in the South Atlantic Ocean approximately 1,500 miles southwest from the coast of South Africa. The irony of the situation is that Tristan da Cunha is a British overseas territory annexed in 1816, and the German raider used it to cover its rendezvous with its supply ship. At this deserted location, *Altmark* was able to replenish the cruiser undetected and unmolested by ships of the Royal Navy. *Trevanion*'s crew was transferred to *Altmark* at this location.

On October 28, once done with replenishment, *Graf Spee* steamed around the Cape of Good Hope into the Indian Ocean. The weather and sea state were, predictably, terrible—strong winds and heavy seas with tremendous swells. Had the cruiser run across any merchant ships, the sea would have been too rough to put a boarding party in the water in a small boat. *Graf Spee* had that part of the ocean to herself, however, and did not sight any merchant shipping on the sea lanes.

As *Graf Spee* moved north, a small tanker traveling in ballast was spotted hugging the African coast. The tanker was *Africa Shell*, owned by the Shell Company of East Africa and displacing only 706 tons. *Africa Shell*'s crew took to their lifeboats and the ship's captain, Patrick Dove, was taken prisoner. The tanker was sent to the bottom. It turned out that Dove was along for the duration of *Graf Spee*'s cruise and was not released until the ship reached Montevideo, Uruguay, a month and a half later. During this time, Dove and Langsdorff became friends, having much in common. Dove published a book titled *I was Graf Spee's Prisoner!*, which detailed his experiences on board the cruiser and his conversations with Langsdorff.

During the ensuing days, *Graf Spee* stopped a number of ships, but all were sailing under the flags of neutral nations. While the ship patrolled the Indian Ocean on November 16, *Graf Spee*'s radio operator intercepted a British transmission to ships in the area warning of the cruiser's presence. The radio decryption, coupled with the lack of targets, was enough for Langsdorff to make the decision to return to the more fruitful waters of the South Atlantic Ocean.

Upon returning to the more fertile hunting grounds in the Atlantic, *Graf Spee* sank the Blue Star Line's 6,347-ton refrigerated cargo liner *Doric Star* approximately five hundred miles off the coast of South Africa. The ship was hauling frozen meat among its cargo, and Langsdorff took enough to refresh provisions on the cruiser. The liner was shelled, and then a torpedo was sent to finish her off.

As the sun rose on December 3, lookouts on *Graf Spee* spotted smoke on the horizon, and the cruiser soon came across the 450-foot-long steamer *Tairoa*, owned by Sir W. G. Armstrong and Whitworth Co. Ltd., of Newcastle-Upon-Tyne, England. The crew of *Tairoa* reported they had heard *Doric Star*'s distress call, and they were trying to get as far away as possible from the estimated position of the Blue Star Line freighter's sinking. German sailors located and removed *Tairoa*'s supply of carbonic acid, needed by *Graf Spee* for its refrigeration plants, which kept the gunpowder magazines at a uniform temperature. Once clear of the freighter, an attempt was made to sink the ship using explosives. When she did not sink, *Graf Spee* shelled her with 150-millimeter cannon fire, and finally sank the stubborn freighter with a torpedo.

After the ship spent the next four days in company with *Altmark* to refuel, replenish, transfer prisoners, and conduct training exercises, *Graf Spee*'s lookouts spotted her last victim as the sun was setting on December 7. The freighter *Streonshalh* was 349 feet long and displaced 3,895 tons, its holds brimming with grain from the plains of Argentina, bound for England. Using her 150-millimeter guns, *Graf Spee* sent *Streonshalh* to the bottom of the Atlantic Ocean and then began to steam westward.

Some men from *Graf Spee* later reported that documents from *Streonshalh* led Langsdorff to believe that a great deal of merchant ship traffic was departing the River Platte from the ports at Buenos Aires, Argentina, on the south side and Montevideo, Uruguay, on the north side of the estuary. The potential of a target-rich

The British steamer *Tairoa* was stopped and boarded by the *Graf Spee* crew, one of nine ships sunk during the *Panzerschiff*'s rampage in the South Atlantic Ocean. After the crew was evacuated and other supplies removed, explosives were placed into the freighter's hull. *Tairoa* did not want to sink, as the explosives failed to scuttle her. After repeated hits from *Graf Spee*'s 150-millimeter cannon, a torpedo was finally able to send her to the bottom. *Author's collection*

SPECIFICATIONS: *ADMIRAL GRAF SPEE*

Length	610 feet 3 inches
Beam	71 feet 0 inches
Draft	24 feet 1 inch
Displacement	14,890 tons
Powerplant	8 MAN diesel engines driving twin screws
Horsepower	52,050 shaft
Crew	30 officers and 920 to 1,040 sailors
Armament	6 11-inch main guns in triple turrets 8 150-millimeter guns 8 21-inch torpedoes
Builder	Reichsmarinewerft Shipyard, Wilhelmshaven, Germany
Class	*Deutschland*-class heavy cruiser (*Panzerschiff*)
Launched	June 30, 1934
Commissioned	January 6, 1936
Destroyed	December 17, 1939

SHIPS SUNK BY *ADMIRAL GRAF SPEE*

Date	Ship	Tonnage
Sept. 30, 1939	*Clement*	5,051
Oct. 8, 1939	*Newton Beech*	4,651
Oct. 7, 1939	*Ashlea*	4,222
Oct. 17, 1939	*Huntsman*	8,196
Oct. 22, 1939	*Trevanion*	5,299
Nov. 15, 1939	*Africa Shell*	706
Dec. 2, 1939	*Doric Star*	6,347
Dec. 3, 1939	*Tairoa*	7,983
Dec. 7, 1939	*Streonshalh*	3,895
	Total tonnage	46,350

environment lured Langsdorff and *Graf Spee* into a battle they ultimately could not win.

HUNTER BECOMES PREY

On the morning of December 13, lookouts onboard *Graf Spee* spotted a pair of masts protruding over the horizon. As they drew closer, the lookouts were able to identify the ship as the Royal Navy's heavy cruiser *Exeter*, followed by the light cruiser *Ajax* and New Zealand's light cruiser *Achilles*. Both the German cruiser and the three Commonwealth ships engaged each other around 6:15 a.m., east of the River Platte estuary. *Graf Spee* fired first, striking *Exeter* with her third salvo. *Exeter*, in turn, launched torpedoes at *Graf Spee*, which the German cruiser outmaneuvered. The battle was on.

Concentrating on *Exeter*, *Graf Spee* made a number of hits while dodging more torpedoes from the stricken British heavy cruiser. By 7:00 a.m., *Exeter* was slowed and listing to port. Langsdorff turned his attention to *Ajax* and *Achilles*, maneuvering

After her encounter with the Royal Navy cruisers *Exeter* and *Ajax* and the New Zealand Navy light cruiser *Achilles*, *Graf Spee* is seen here in Montevideo Harbor, Uruguay, with heavy battle damage. Note the large shell hole on the port side, aft of the anchors. *Author's collection*

Graf Spee's Arado floatplane was completely burned, and damage from British and New Zealand shells can be seen on the ship's side. *Author's collection*

Captain Langsdorff, at left, committed suicide two days after scuttling his ship. He was buried with full military honors. The surviving men from *Graf Spee* owed their lives to the captain. *Author's collection*

to avoid presenting an easy target for torpedoes from the pair of enemy light cruisers.

At 7:20 a.m., both *Ajax* and *Achilles* turned and fired at *Graf Spee* simultaneously, scoring a number of hits on the German cruiser. *Graf Spee* took *Ajax* under fire, putting both of her main batteries out of commission. *Ajax* launched a spread of four torpedoes, but *Graf Spee* turned to present its stern and the four "fish" missed the now-narrow target. *Ajax* and *Achilles* closed *Graf Spee* again, but all three ships, independently, turned and broke off the engagement. The Battle for the River Platte was over by 8:00 a.m.

The two Commonwealth light cruisers took up station following *Graf Spee* as she headed for Montevideo, and *Exeter* turned to limp to the British Falkland Islands, where she could make repairs. Onboard *Graf Spee*, thirty-six had been killed with another sixty wounded. Langsdorff surveyed damage to his ship. He found that the fresh water plant was seriously damaged, the galleys were out of commission, there was a six-foot-square hole in the bow, and the engine oiling and fuel purification systems were severely damaged. Moreover, one of the ship's 150-millimeter cannon of the secondary armament was completely destroyed, and his engines were tired after six months at sea, with cracked pistons and cracked cylinders that were in desperate need of repair.

Surveying the heavy cruiser's offensive capabilities, he found that he had only 186 shells for his eleven-inch guns, seven torpedoes, and 2,920 shells for his 105-millimeter anti-aircraft guns. German naval planners severely underestimated the amount of ammunition required for a ship-to-ship battle, and at the most, Langsdorff could engage targets for about twenty minutes. Then *Graf Spee* would be out of ammunition and forced to surrender or be annihilated by superior Royal Navy forces.

Langsdorff entered Montevideo Harbor in neutral Uruguay to effect repairs to the ship, with the intention of steaming back to Germany for more permanent work. Limited to forty-eight hours in port by the Uruguayans, Langsdorff appealed for assistance from nearby German ships because the local marine repair yards refused to aid the heavily damaged ship. The heavy cruiser's captain felt he needed thirty days to make sufficient repairs, while a German shipwright from a civilian freighter thought most of it could be accomplished in two weeks' time—time Langsdorff did not have. The following day, Langsdorff addressed his crew and told them he would not sacrifice them trying to do battle against a superior naval force.

While *Graf Spee* was in Montevideo Harbor, the British were rushing more warships to the area to confront the German heavy cruiser when she headed out to sea. Those forces were not expected to arrive off the River Platte estuary until

Faced with an unwinnable situation, having only 186 shells for its eleven-inch main guns and 2,920 shells for the 105-millimeter secondary armament, Capt. Hans Langsdorff surmised that he only had enough ammunition for twenty minutes of battle. Rather than sacrifice his ship and crew in a pointless engagement, he sailed the ship out of Montevideo Harbor with a skeleton crew and scuttled her. *Author's collection*

As *Graf Spee* burned at the mouth of the River Plate on December 17, 1939, more than one thousand survivors from the ship's crew were taken to Buenos Aires, Argentina, where they were interned for the duration of the war. *Author's collection*

December 19. That did not, however, stop the British from putting out false information that the aircraft carrier *Ark Royal* and heavy cruiser *Renown* were already patrolling in the vicinity.

Facing what he believed were overwhelming odds, Langsdorff put a skeleton crew on board the cruiser and sailed out of Montevideo Harbor on December 17. Once off the coast, he scuttled the ship. *Graf Spee*'s crew of more than a thousand was taken to Buenos Aires, Argentina, where they would be interned for the duration of the war. On December 19, Langsdorff committed suicide. He was buried in Buenos Aires with full military honors, and a number of British naval officers attended the ceremony.

DIVING THE *GRAF SPEE*

"For Uruguay, *Graf Spee* is a story that is still being told even today, because it was the first time my country was touched by a war, by a foreign war, like the Second World War," said professional diver Hector Bado, who has worked to salvage the wreck of *Admiral Graf Spee*. "I think one very interesting aspect about the story is not the ship itself, but its captain. Captain Langsdorff should be remembered for his gallantry, and he never hurt a merchant seaman when he sank those nine ships. He decided not to fight the British, not because he was a coward or anything like that, but because the situation was hopeless."

As a young man, Bado was taught to dive by his father, and the two spent many hours underwater, exploring the area around the River Platte estuary. Bado said:

A solitary buoy marks the final resting place of the *Graf Spee*. *Hector Bado*

I was very interested in shipwrecks. Around 1985 I started diving as a professional career, and in the course of my work we found a lot of shipwrecks there in Uruguay, from the seventeenth century on.

One thing takes you to the next, and in 1996 we did a job for the Uruguayan navy. They were searching for a lost anchor and chain from a freighter. That day we were very close to the wreck of the *Graf Spee*, so I suggested to the navy divers that we have a look at the German warship. We were using side-scan sonar at that time and made a couple of passes close to the wreck to see what condition it was in because it's not visible right now. When there is low tide, only one foot of the main forward turret is visible.

To our surprise we found that the wreck was in pretty good shape. We made a note of the condition and continued our work. The next year, we were working with a university archaeology program, and they called me from England to ask me to participate in a TV documentary about the HMS *Agamemnon*, which was the favorite ship of Lord Nelson. We found that ship in Uruguayan waters as well. While working on the HMS *Agamemnon* documentary, I proposed that they make a second production on *Graf Spee*—kind of a two-for-one deal.

The production company gave us the go-ahead, and we did some research on the *Graf Spee* wreck by diving her. We were very surprised again about the condition of the wreck. For these trips we brought new equipment, new side-scan sonars, and we were able to get the help of the Uruguayan government. We raised one of the one-hundred-and-fifty-millimeter guns from the secondary armament. That was in 1999. It was a big success.

CONDITION OF THE WRECK

Graf Spee sits 4.3 miles southwest of Montevideo's port in about thirty feet of water. There's a strong current, and the water is virtually black. It is a very dangerous place to dive. When the current, slows, particulate matter in the water settles, and occasionally visibility improves from one foot to twelve or fifteen feet.

The ship rests at a 65° angle to port, and currents on the starboard side stack sediment against the hull while those eddying on the port side have carved out the seabed to a depth of approximately thirty-six feet. Bado recalls:

> One day in particular we had fifteen to eighteen feet of visibility underwater and we were able to see part of the main bridge, parts of the funnel, and, although it was listing and touching the bottom, parts of the mainmast; all kinds of guns are scattered on the bottom. It's very interesting. We never expected the structure was going to be in such good condition.
>
> We do not enter the wreck because it's very dangerous. I'm the only one from our group who did, and I was trying to penetrate the forward turret. I went inside about thirty feet and I thought to myself, "Enough." There was no way to see anything. You can get lost inside the wreck and that would be the end of you. After that experience, I decided we are not going to penetrate the wreck. As I recall, in 1941, two British divers were trying to recover the rangefinder and they drowned. It's not a very safe place to dive, and getting inside the wreck is a crazy idea.

RAISING THE WRECK?

In 2003, Bado approached the Urguayan government about salvaging the ship. With permission, he and his team recovered the thirty-one-foot-wide rangefinder from the conning tower of the ship. Although many other German naval vessels were equipped with this size of rangefinder—*Bismarck*, *Scharnhorst*, *Gneisenau*, and *Prinz Eugen*—this is the only example on dry land that historians can inspect. After the rangefinder was recovered, it was restored and put on display in Montevideo.

Then there was talk about raising the stern of the ship. When *Graf Spee* was scuttled, sailors placed three charges inside the ship. Two of the charges exploded and a third failed. One of the exploding charges was placed near the rear turret and the other in the engine room. This combination of explosions blew off seventy feet

In 2003, Hector Bado and his team recovered one of the range finders from the *Graf Spee* on behalf of the Uruguayan government. The twenty-seven-ton, thirty-one-foot-wide range finder was restored and is now displayed at the Uruguayan Naval Museum in Montevideo. *Hector Bado*

of the ship's stern. All of the deck equipment remains on this section, including the torpedo launching tubes; however, efforts to raise the stern section were abandoned due to a lack of funds. At the time, it was estimated that the cost to raise the entire ship would be in the neighborhood of $30 million plus an unknown amount to conserve her. Even attempting to conserve the stern section would have been extremely costly.

Bado reports that the bow of *Graf Spee* is in pretty bad shape. In 1942, the Uruguayan Port Authority attempted to recover the anchor chain and anchors from the ship but only succeeded in slicing up the bow area. In addition, around that time, British Royal Navy intelligence experts went aboard and salvaged items of interest. Later that same year, a floating crane was working the site when it capsized, killing fourteen men. After the accident, most people stayed away from the wreck.

SALVAGING THE STERN EAGLE

Three months after *Graf Spee* was laid down, the Nazi Party took over the German government. The party's symbol was an eagle with outstretched wings grasping a *Hakenkreuz*, or swastika, in a laurel wreath. When *Graf Spee* was built, a bronze Nazi eagle with a nine-foot wingspan was affixed to the stern of the ship—certainly the most recognizable relic from the cruiser, yet one that stirred emotions in many people.

Most in the diving community believed the eagle had been removed long ago. To his surprise, Bado encountered an old diver friend in Uruguay who knew of his interest in *Graf Spee*. The diver told Bado that a group was trying to salvage the eagle and that it was still affixed to the ship. "In 2006, I decided to visit the wreck and try to find it. I actually found it in a couple of hours. The eagle was there," Bado said. "Then I went to the naval authorities in Uruguay and talked to the commander-in-chief of the navy. We talked in total secret because I knew from the very beginning that it was going to be a problem because Nazi relics are very controversial items.

So I asked what should be done because if we don't salvage the eagle, it's going to be stolen. The navy gave me permission to proceed, but they said nobody could know what we were up to, and we could only take a minimum of personnel."

For eighteen days, Bado and two diver friends worked to unscrew the 145 bolts that held the eagle to *Graf Spee*'s stern. "We were working eight hours a day. It was painful," he said.

> When you work on the *Graf Spee*, you are working in pitch-black waters. For example, if you turn on your underwater flashlight and you put the flashlight just in front of your diving mask, you'll only see something glaring. You will not see the light itself. It's like liquid mud down there. So in some way we were risking

For the 2006 expedition to *Graf Spee*, this barge was used to recover the bronze stern eagle from the ship. *Hector Bado collection*

The stern eagle is lifted aboard the barge. Divers worked in near-zero visibility to unbolt the Nazi symbol from the ship before pirate salvors could steal and sell it to be recycled for its metal content or other nefarious reasons. *Hector Bado collection*

our lives in that hole trying to get the eagle. Fortunately, everything went well and we salvaged the eagle.

When we were taking the eagle to port, I decided to cover the swastika with a yellow tarp because we wanted to be respectful, and we didn't know how this symbol would be seen by the Jewish community in Uruguay, or anybody else for that matter. We decided to talk to all the parties involved in this situation, and then we would make a decision as to the eagle's final fate. We talked to the president of the Jewish community in Uruguay together with the navy admirals. We all decided that history is history and you cannot hide history. So the eagle was finally put on display for three weeks in a hotel in Montevideo. More than thirty thousand people went there to see it. After that, the eagle was put in a box, and to this day today it is still under guard.

We are now fighting about our rights because half of that eagle belongs to us, according to the contract that we signed with the Uruguayan government. But in the middle of all this mess, the German government claimed that the ship was theirs. But we know that the ship was sold in 1940 by the German ambassador to a British spy for £16,000 [roughly $24,500 in today's dollars]. We have the original documents. However, Germany is the main weapons supplier to the Uruguayan navy. So we are fighting to make our end of the contract call for the eagle to be delivered and displayed in a museum. It's taking a little bit longer than we expected because we've been dealing with these problems since 2006 and they're still not resolved.

I don't think that the eagle is a very important part of Uruguayan history, but I do think that the eagle should go to a museum, somewhere. Obviously the Germans don't want anything to do with it. I think that the eagle should be shown somewhere, either in Uruguay or in Europe, or possibly in Great Britain, at a suitable location, such as the Imperial War Museum in London. But it's an item that is so impressive, and in some ways it's important because it's the only one surviving in the world. Only five were cast and this is the only one that survived the war. So it's a pretty impressive part of Second World War history, mainly because it belonged to a ship of importance, the *Graf Spee*, which was the flagship of the German navy. The *Graf Spee* made a big mess sinking all these merchant ships. And after that is a tale of the gallantry of the captain. There is a very rich story about the *Graf Spee*, and it involves a lot of different aspects that may capture the public's attention.

FUTURE HOPES

For now, Bado cannot get close to *Graf Spee* or any of the 350 other shipwrecks he's found at the mouth of the River Platte. Uruguayan laws currently restrict the salvage of ships in the nation's coastal waters. It will take a change in government to change the laws, and the first opportunity for this to occur won't happen until 2016 at the earliest. Bado said, "Hopefully, in the coming years, we will be able to return to recover *Graf Spee* and other shipwrecks of major importance for our country."

The bronze eagle is more than six feet tall, with a wingspan of nine feet. *Hector Bado collection*

In spite of the abhorrent symbolism represented by the swastika, *Graf Spee*'s eagle is a true representation of the sculptor's art. It is finely detailed down to the breast feathers. Diver Sergio Pronczuk, left, and Hector Bado pose with it on the barge above the shipwreck. The eagle was displayed in Montevideo, as the ship is part of Uruguayan culture; it has since been crated and stored in a Uruguayan Navy secure warehouse. *Hector Bado collection*

U-550

HUNTING THE SUB CODE-NAMED "RED GEORGE"

U-BOAT CREWS WERE A THREAT TO EVERY Allied ship on the sea until the Germans unconditionally surrendered on May 7, 1945. German efforts to starve England and Russia by denying access to the sea lanes to transport food, fuel, and military equipment were only hampered when the convoy system from the United States to the United Kingdom was finally enacted in spring 1942.

By April 1944, convoys of tankers and freighters sailing from the US East Coast were departing for the United Kingdom and Russia. They typically comprised more than 110 ships each, plus armed escort vessels. This "ocean bridge," which brought food, fuel, and armament to America's allies, presented a rich target environment for German navy (*Kriegsmarine*) submarines, but any attack was made at substantial risk. At this point in the war, the Allies had built up significant antisubmarine warfare forces, and the area known as the mid-Atlantic gap, where submarines could, for a time, operate without fear of long-range Allied patrol planes, was now closed. The Allied antisubmarine noose was beginning to wrap around the German U-boat force.

As March 1944 turned into April, the US Navy's Eastern Sea Frontier began tracking a pair of U-boats as they moved through the North Atlantic headed for Canadian and then American coastal waters. The two submarines were given the designations of "Red Fox" and "Red George," and their positions were followed closely as they transited from the north Atlantic Ocean through the Flemish Cap (an area about 350 miles east of St. John's, Newfoundland—47° north, 45° west) and on toward the United States. The subs moved south on the surface, hoping to catch a convoy or an unescorted ship venturing between ports.

America's industrial might produced hundreds of destroyers, destroyer escorts, and smaller versions of aircraft carriers known as CVEs, or carrier escorts, affectionately known as "jeep carriers." For example, *Bogue*-class escort carriers were built on the US Maritime Commission's Type C3 hull, essentially a freighter hull with a flight deck. Three shipyards (Seattle-Tacoma Shipbuilding Corporation in Washington;

Ingalls Shipbuilding in Pascagoula, Mississippi; and Western Pipe and Steel in South San Francisco) turned out forty-five *Bogue*-class escort carriers beginning in 1942. In all, US shipyards built 122 escort carriers of all classes by summer 1945.

Each *Bogue*-class escort carrier was 496 feet long overall with a beam of 69 feet 6 inches and a draft of 26 feet, displacing 16,620 tons. Top speed was eighteen knots, or twenty-one miles per hour. The flight deck was 439 feet long and 70 feet wide, not very big when one considers that a Grumman TBF Avenger torpedo bomber's wingspan was 54 feet 2 inches. Both the US Navy and the Royal Navy operated escort carriers, and their typical complement of aircraft was twelve F4F Wildcat fighters and nine Avenger torpedo bombers for US ships, or almost the same number of Supermarine Seafire fighters and Fairey Swordfish bombers for Royal Navy escort carriers.

Working as "hunter-killer" task groups, each escort carrier was paired with three or four destroyers. The escort carrier's aircraft could range far and wide in search of U-boats sitting on the surface or cruising submerged with their periscopes exposed. The aircraft would attack, mark the U-boat's last location, then orbit until the destroyers arrived on scene. It was these hunter-killer task groups that closed the mid-Atlantic gap and slowed the number of ships that fell prey to U-boats.

The tactical doctrine of the hunter-killer group was to exploit the U-boats' weaknesses. In today's terminology, the German navy "micromanaged" its U-boats by having them report to headquarters every day. Allied direction finding—known as "HD," "Huff-Duff," or "RDF," for radio direction finding— was able to triangulate the submarine's position and then dispatch a hunter-killer group to destroy the U-boat.

In addition to the threat of their positions being exposed through RDF, the U-boats had four other vulnerabilities; any one or a combination could present the Allies with the opportunity to kill a U-boat. First, Allied antisubmarine forces could simply calculate the range from a submarine's last reported position to a point where the U-boat's battery charge would be exhausted and the vessel forced to surface. The second variable to consider was the point where the U-boat would run out of fresh air; the submarines had to be ventilated regularly to exhaust fumes from the batteries and other noxious odors.

Third was the maximum endurance of the crew—how long they could stay down physically and how much mental punishment they could take if the boat was being depth charged. The final variable was speed. The U-boat's maximum speed surfaced and its maximum speed underwater were known; thus, an estimated position could be derived from these factors as well.

All of these calculations combined to form a search box, and if the sub's last direction of travel was known, that further reduced the search area. Position a patrolling aircraft above these points and wait for the U-boat to surface; the aircraft could strafe the boat, fire rockets, or drop depth charges to damage or kill the boat. It also gave the pursuing Allied forces a fresh position from which to continue tracking the sub.

The pursued U-boat's death knell sounded when an escort carrier's aircraft dropped sonobouys, or if the sub was caught on the surface and attacked with rockets, machine gun fire, or depth charges. Once an aircraft expended its ordnance, it could loiter in the area until supporting destroyers, destroyer escorts, or corvettes (depending upon configuration of the task group) arrived to continue the hunt for the U-boat.

The submarine, on the other hand, relied on stealth tactics. It could often submerge before it was sighted by an aircraft; it could detect hostile radar at ranges greater than most Allied airborne radar could detect the submarine; and a U-boat could dive deep and sit motionless under the water. During many great black-and-white submarine war movies from the 1950s and 1960s, the hero captain would command, "Rig for silent running!" It made for a dramatic movie line, and it was also a deadly serious configuration for Allied and Axis submarines alike. Depending upon temperature layers in the sea, sonobouys and other surface-based sonar equipment could not detect a submarine underneath various thermoclines, and if the boat was completely still, acoustic underwater microphones could not detect machinery noises, either.

SCRATCH RED FOX

Red Fox and Red George were headed for an area on alert for the subs with Task Group 21.15, built around the escort carrier *Croatan* (CVE-25) and her destroyer escorts, already on antisubmarine patrol. On April 2, *Croatan*'s aircraft and its destroyers had attacked a submarine at 40°20' north latitude, 60°33' west longitude. Contact was lost, but soon destroyer *Carmick* (DD-493) regained contact, making three separate depth charge attacks. Between the second and third attacks, the U-boat fired a torpedo at *Carmick*, but it missed. Both Red Fox and Red George were sailing parallel to each other, approximately forty miles apart, and after the confrontation with *Croatan*, no contact was made with either submarine for the next five days.

At 3:25 a.m. on the morning of April 7, *Croatan* launched a pair of TBM-1C Avenger torpedo bombers from Composite Squadron 42 (VC-42) to hunt for the subs. That morning, *Croatan*'s air operations officer recorded the day's flying conditions as "Undesirable in the morning becoming bad in the afternoon. Surface winds 25 knots at 2 a.m. Sea state—Rough." Perfect weather to stay in your rack, not to wander around on a pitching flight deck in the dark, let alone go flying.

Launched from *Croatan*'s pitching deck into the dark early morning sky were Lt. Wilburt A. Lyons, flying Grumman TBM-1C Avenger Navy serial number (referred to as Bureau Number, or BuNo) 25484, which wore number "18" on the fuselage; and Lt. j.g. C. G. Mabry Jr., in TBM-1C BuNo 25507 (aircraft "21").

Lieutenant Lyons made contact with a surfaced submarine at 40°33' latitude, 62°22' longitude. It was a moonless night, and Lyons's radioman, F. C. McKee, picked up the sub on radar at a distance of four and a half miles. Passing over the target, Lyons dropped a sonobuoy pattern and a dye marker, but it was too dark to fire his rockets or machine guns at the vessel. The target was quickly confirmed to be a German U-boat; as the Avenger passed overhead, the submarine's antiaircraft gunners began putting holes in its underside. While the aircraft flew out of

Left: When *Pan Pennsylvania* was launched at the Welding Shipyards in Norfolk, Virginia, in 1943, she was one of the largest tankers afloat at 11,016 gross registered tons, measuring 515 feet 11 inches long with a beam of 70 feet. Her two steam turbines could move this mass through the water at fourteen knots. On April 15, 1944, the ship was carrying a cargo of 140,000 barrels of aviation fuel. While forming into a convoy, *Pan Pennsylvania* was struck in the port side, aft, by a torpedo from the U-550. *Author's collection*

SPECIFICATIONS: *U-856* "RED FOX"

Type	IXC/40
Length	251.8 feet
Beam	22.5 feet
Draft	15.3 feet
Speed	19 knots surfaced, 7 knots submerged
Range	13,850 miles at 10 knots
Torpedoes	22 (4 bow tubes, and 2 stern)
Deck gun	1 105-millimeter/45-caliber
Crew	55
Maximum depth	775 feet
Builder	AG Weser, Bremen, Germany
Laid down	October 31, 1942
Launched	May 11, 1943
Commissioned	August 19, 1943
War Patrols	1 (March 1–April 7, 1944)
Ships Sunk	None
Lost	April 7, 1944

range, the U-boat dived under the surface, no doubt hoping to escape further detection. Mabry soon arrived and the two Avengers circled the area listening to the sonobouys. They failed to pick up the submarine, but they continued to orbit the area.

A hunter-killer group of two F4Fs and two TBMs was dispatched at 6:00 a.m. to take up the hunt. They, too, could not get a bearing on the U-boat.

At 7:10 a.m., destroyer *Boyle* (DD-600) arrived to assist in the search. One hour after arriving on scene, *Boyle* made contact approximately a thousand yards away. The destroyer ran down the contact but did not drop depth charges for fear that the explosions might interfere with the operations of the patrolling aircraft.

Boyle turned and ran down the contact again, this time firing a pattern of eleven depth charges set to explode between two hundred and three hundred feet. Before the charges reached their set depths, a periscope was sighted five hundred yards astern of the ship. The periscope disappeared, the depth charges went off, and sonar contact with the submarine was lost.

Soon thereafter, *Boyle* was joined by the destroyers that made up Task Unit 27.6.1—*Parker* (DD-604), *McLanahan* (DD-615), and *Laub* (DD-613). *Frost* (DE-144) and *Huse* (DE-145) stayed behind to protect *Croatan* but were ready to join in the fray if required. The four destroyers began a box search for the as-yet unidentified U-boat at 9:45 a.m. The hunt stretched into the afternoon, ending at

SPECIFICATIONS: *U-550* "RED GEORGE"

Type	IXC/40
Length	251.8 feet
Beam	22.5 feet
Draft	15.3 feet
Speed	19 knots surfaced, 7 knots submerged
Range	13,850 miles at 10 knots
Torpedoes	22 (4 bow tubes, 2 stern)
Deck gun	1 105-millimeter/45-caliber
Crew	56
Maximum Depth	775 feet
Builder	Deutsche Werft AG, Hamburg, Germany
Laid-down	October 2, 1942
Launched	May 12, 1943
Commissioned	July 28, 1943
War Patrols	1 (February 1–April 16, 1944)
Ships Sunk	1 ship of 11,017 gross registered tons
Lost	April 16, 1944

3:00 p.m. Minutes later, *Frost*, *Parker*, *McLanahan*, and *Laub* began a retiring search to the west while *Huse* and *Boyle*, along with the destroyers composing Task Unit 27.6.2—*Champlin* (DD-601), *Nields* (DD-616), and *Ordronaux* (DD-617)—headed east, their sonars pinging into the deep in the hope that one of the pings would reveal the U-boat's location.

At 4:42 p.m., *Champlin* made contact at 1,600 yards but lost it while closing to attack. Contact was regained and lost again; however, this time a smoke pot was dropped over the estimated location of the sub. *Huse*, which had gone to assist *Champlin*, made contact, lost contact, subsequently regained contact, and at 5:14 p.m. fired a Hedgehog depth-bomb pattern. Two muffled explosions were heard, which indicated a direct hit, as Hedgehogs were detonated by contact rather than by depth pressure like a standard depth charge.

After the destroyers stood off to let the water settle, at 5:34 p.m., *Champlin* ran in and dropped nine depth charges set deep, to explode between five hundred and six hundred feet. Nothing happened for six minutes. Everyone on deck was watching the patch of ocean where the depth charges had disturbed the surface. As the sea settled down, destroyer *Huse* regained contact, but lost it again as she turned to attack. Many a World War II sailor has said that antisubmarine warfare was an often-frustrating, high-stakes game that took patience on the part of the pursuers.

Less than forty-five days before the tanker was sent to the bottom of the Atlantic Ocean by *U-550*, the Coast Guard took this photo of *Pan Pennsylvania* at sea. The ship sailed with a crew of fifty plus thirty-one armed guards who manned one five-inch and one three-inch guns (bow and stern) as well as eight twenty-millimeter antiaircraft guns. *US Coast Guard*

At 5:45, *Huse* regained contact and dropped eight depth charges that essentially blew the U-boat to the surface. *Huse* and *Champlin* immediately began firing upon the submarine with twenty-millimeter cannon. The submarine returned fire as both destroyers attempted to ram the sub. Moving in, sailors on *Huse* could see survivors in the water and the destroyer hove to, as it appeared the order to abandon the U-boat had been given.

At 6:15 p.m., destroyer *Champlin* reported the sub sinking, eventually sliding under the sea at 40°18' north latitude, 62°22' west longitude. Destroyers *Nields* and *Ordronaux* fished the captain, two officers, and twenty-five crewmen from the ocean. They confirmed the identity of the submarine as the Type IXC/40 *U-856*, which was on its first war patrol under the command of *Oberleutnant* Friedrich Wittenberg. He was only twenty-five years old.

IN PURSUIT OF RED GEORGE

Red George was a lucky boat. On February 22, 1944, at the start of its first war patrol, a PBY Catalina flying boat from the Royal Canadian Air Force's 162 Squadron (plane "S"), under the command of Flying Officer C. C. Cunningham, surprised her as she lay on the surface. Patrolling south of Iceland, Cunningham was able to straddle the boat with four depth charges while his gunners strafed the submarine's decks. Two of the German U-boat's crew were killed in the attack.

Then Red George made a lucky escape from destroyer *Carmick*'s April 2 attack. Although the U-boat attempted to end the engagement by firing a torpedo at *Carmick*, it was able to escape detection, thus ending the pursuit.

Red George continued sailing south. As it was later learned, she was a very potent Type IXC/40 U-boat under the command of *Kapitänleutnant** Klaus Hänert. The Type IXC/40 submarines were faster and had an increased range over standard

* *Kapitänleutnant* (*Kptlt.*): equivalent to the US Navy rank of lieutenant commander

Type IXCs. The Type IXC/40 was designed as an ocean-cruising submarine with a range of 13,850 nautical miles when cruising on the surface at ten knots, or sixty-three nautical miles at four knots while submerged. This gave the Type IXC/40 an approximately fifty-day cruising capability before it had to be refueled and re-provisioned. On the surface, these U-boats had a top speed of 18.3 knots per hour (21.1 miles per hour), and maximum speed submerged was 7.3 knots per hour (8.4 miles per hour). Eighty-seven were built.

During the afternoon of April 15, 1944, twenty-eight merchant ships and six destroyer escorts left New York Harbor bound for the United Kingdom. Known as Convoy CU-21, the ships were traveling from oil-rich Curaçao to Liverpool, United Kingdom—thus the CU designation. Escort ships for the voyage were six destroyers: *Gandy* (DE-764), *Harveson* (DE-316), *Joyce* (DE-317), *Kirkpatrick* (DE-318), *Peterson* (DE-152), and *Poole* (DE-151). *Joyce* and *Peterson* were two of the thirty destroyer escorts operated by US Coast Guard crews. Each was 306 feet long with a beam of 36 feet 7 inches and a top speed of 21.5 knots. They were a formidable antisubmarine platform, equipped with eight depth-charge projectors, two depth-charge racks at the stern, one triple torpedo-tube mount, three three-inch/fifty-caliber guns, one twin forty-millimeter gun, and eight twenty-millimeter cannon. This armament combination enabled the DEs to blow a submarine to the surface with depth charges and send it back to the bottom with cannon fire or a torpedo.

Sailing weather was poor that April afternoon, and the convoy was directed to steer south of its intended route to allow for an inbound group of ships to pass.

Destroyer Escort *Joyce* (DE-317) and its sister-ship *Peterson* (DE-152) were both manned by Coast Guard crews and were assigned to escort Convoy CU-21 to Liverpool, England. *Joyce* picked up survivors from the *Pan Pennsylvania* and as she got underway, her sonar detected the *U-550*. Notice the depth charge racks on the stern of the destroyer escort. *US Coast Guard*

With the change of course, poor weather, and the darkness of night coming on, it was decided not to form into the convoy until the light of the next morning. That evening, the ships traveled in two columns of fourteen ships, each making fourteen knots. Three destroyer escorts were patrolling on each side of the string of ships as they moved eastward.

In the dark of night, at approximately 10:30 p.m. local time, the US-flagged tanker *Sag Harbor* and the Honduran-flagged *Aztec* collided. Destroyer escort *Peterson* was dispatched to shepherd the two damaged ships back to New York Harbor.

While *Peterson* escorted the pair of ships, the Eastern Sea Frontier's Submarine Tracking Section gave an update on its estimated position of Red George. Their estimate put Red George at 38°00' north, 71°00' west in the general area "600 miles east of Cape May," New Jersey. The Submarine Tracking Section's report put Red George in deep water off the continental shelf, not yet a threat to Convoy CU-21.

U-550 sinks stern first as her crew abandons ship. Based on the angle of this photo, the destroyer in the foreground is most likely *Gandy* (DE-764), which had just rammed the submarine's stern section. *US Coast Guard*

Steaming east at 8:05 a.m., two hundred miles from New York Harbor, the ships of CU-21 were attempting to form into convoy columns. Moving at fourteen knots, the 11,017-ton tanker *Pan Pennsylvania* suddenly shook violently. A torpedo from Red George had found her, impacting on the port side, aft, between the No. 7 and No. 8 tanks. A fire started in the boiler room, the steering gear was disabled, and the ship began taking on water. The ruptured tanks began releasing some of the tanker's 140,000 barrels of gasoline onto the ocean's surface, and the highly volatile fuel quickly surrounded the ship as she slowed in the water. Still making way, a number of men launched a lifeboat while the tanker was moving; many of them perished

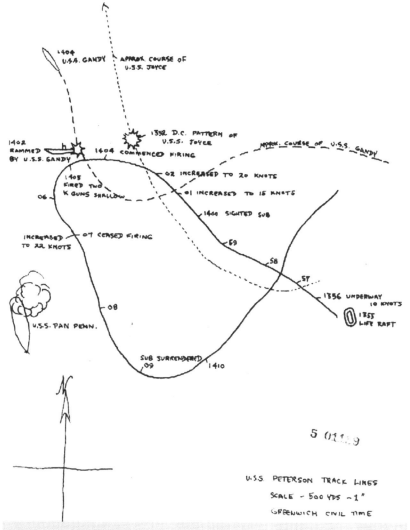

A map of the engagement with *U-550* made on board *Peterson*, showing where the destroyer escort picked up survivors from *Pan Pennsylvania*, her speeds as she pursued the U-boat, where she fired two salvos of depth charges from her K-guns, and her proximity to the other ships attempting to sink the submarine. *US Navy*

when the lifeboat overturned, and others drowned while awaiting rescue. The master of the vessel, Capt. Delmar M. Leidy, was able to stop others from rushing to launch a second lifeboat as he made his way to the engine room to stop the ship's forward movement and check the hull damage firsthand.

Destroyer escort *Gandy* began searching for the submarine, which (as was later learned) was hiding very close to, and sometimes beneath, the torpedoed tanker. Returning to the scene of the attack to cover *Joyce* while she fished the survivors from the water, lookouts on *Gandy* reported that a torpedo had missed the destroyer. Fifty-six of the eighty-one men aboard *Pan Pennsylvania* were rescued.

Peterson, returning from its side trip back to New York, was ordered to assist *Joyce* in pulling survivors from the water. At 9:45 a.m., *Joyce* finished picking up survivors and began moving to cover *Peterson*. As *Joyce* got underway, her sonar operators made contact with Red George and dropped a depth-charge pattern. While the water settled, *Peterson* began steaming. Now all three escorts were moving, hunting for Red George.

At 10:00 a.m., Red George's bow pierced the surface at a 30° nose-up angle, about six hundred yards from *Gandy*. The destroyer escort began firing at the submarine and increased speed to ram it. *Gandy* struck Red George approximately twenty-five to thirty feet from her stern, and the destroyer escort turned hard left to prevent its own propellers from striking the crippled submarine. Cannon fire from the destroyer escorts that sailed over the submarine set the fuel from *Pan Pennsylvania* on fire, engulfing the listing tanker.

The radio operator on board *Gandy* heard a transmission in German, and it was interpreted as an order to abandon ship. With that, *Gandy* ceased fire; once the fusillade stopped, the submarine reportedly began firing back. Both *Gandy* and *Peterson* opened up again with twenty-millimeter fire, which quickly silenced Red George's gunners, and the men of the submarine immediately abandoned ship. *Peterson* fired two Mk 9 depth charges from its starboard K-gun depth-charge thrower set to explode at thirty feet. Red George had only moments to live. At 10:34, the U-boat slid under the Atlantic Ocean, stern first.

Red George had revealed itself, and she was no match for the superior force of four destroyer escorts. Red Fox was dead.

Destroyer escort *Joyce* began picking up the submarine's surviving crew. Thirteen survivors from Red George were fished out of the ocean, one dying aboard ship, while the remaining forty-three were lost. Ships passing through the area later picked up three bodies. *Kapitänleutnant* Klaus Hänert was among the U-boat's surviving crew, and he confirmed the submarine's identity as *U-550*.

ON THE BOTTOM: *PAN PENNSYLVANIA* AND *U-550*

As scuba gear improved between the 1970s and the 1990s, sport divers were able to safely go to greater and greater depths, making many historic World War II–era shipwrecks available to explore. In addition to *U-550*, the East Coast is home to the wrecks of *U-352*, *U-701*, *U-853*, *U-869*, and others, many safely within reach of experienced sport divers.

Postwar accounting of the fates of Germany's U-boats gave a good tally of the submarines sunk both during and after the war; however, many of their exact locations were not known. Over the years, a number of groups had pursued *U-550*, but the U-boat eluded all searchers. After an exhaustive effort to research the events surrounding the loss of *U-550*, on July 23, 2012, a team headed by Joe Mazraani announced it had solved the mystery of the U-boat's final resting place.

"I started diving in the 1990s," said Mazraani. "I was already interested in history and always interested in finding shipwrecks, and I wanted to find a wreck where no one had ever been. When *U-869* was officially confirmed, it was big news, and I became interested in *U-550*, among other ships."

Before finding the *U-550* wreck site, a number of divers had visited the wreck of her victim, *Pan Pennsylvania*. "When they found *Pan Pennsylvania* back in 1994, they only dove it a couple of times," Mazraani said. "The visibility was poor, and the previous divers were not in a position on the wreck to determine whether the ship was intact or not. Having located the forward section, they thought the ship was whole. When we were doing the search for *U-550*, our sonar graphed a very large target, and we started to wonder what it could be. Given the location, we initially thought that *Pan Pennsylvania* might have broken in half. That turned out to be the case, because when we dove it in 2013, we confirmed that it was the stern of *Pan Pennsylvania* from the size and the engine."

Pan Pennsylvania's bow section floated for two days before it was intentionally sunk because it was a hazard to navigation. The two halves of the ship are resting twenty miles apart; thus, Mazraani and his team believe the ship broke in two shortly after the torpedo attack, most likely when it capsized. Once the weight of the heavy stern section and all of its machinery spaces were gone, the buoyant bow section, filled with gasoline, continued to be pushed along with the currents until sent to the bottom.

The expedition to find *U-550* set out aboard Mazraani's forty-five-foot dive boat *Tenacious* in 2011, 2012, and 2013. *Tenacious* can accommodate seven divers and is usually docked at Point Pleasant Beach, New Jersey. Because the search area is off Nantucket, Mazraani repositioned the boat to Montauk Point, New York, for both expeditions. From Montauk Point, it's a ten- to twelve-hour boat ride to *U-550* dive site.

A beaming Joe Mazraani onboard his dive boat *Tenacious* after a dive to *U-550*. Mazraani gathered the team of divers and the sonar expert who found the long-missing German U-boat. *Bradley Sheard*

Mazraani, a criminal defense lawyer by trade, employed sonar technician Garry Kozak from Edge Tech in West Wareham, Massachusetts, for both the expedition to map the sections of *Pan Pennsylvania* and to find *U-550*. The other members of Mazraani's team included tugboat captain Eric Takajian, commercial diver Anthony Tedeschi, and divers Steve Gatto (an electrician), Tom Packer (an HVAC contractor), and Bradley Sheard (an aeronautical engineer, author, and underwater photographer). Also part of the team was adventurer and author Randall Peffer, who wrote a book about the expedition titled *Diving the Last U-Boat*, set to be published in 2015.

"The six deep-wreck divers on this expedition welcomed me to their band of brothers ten months after their discovery of *U-550* in the murky light more than three hundred feet beneath the surface of the North Atlantic," said Peffer. "As *Diving the Last U-Boat* took shape, I joined them on numerous dive trips to remote deep wrecks near the edge of the continental shelf off New York, New Jersey, and New England. We stood watches together aboard *Tenacious* and endured long trips in bone-jarring seas. We navigated through the fog and ship traffic. We held our breath together one day sixty-five miles offshore when we had two divers down on the wreck of the freighter *Bidevind* and a monster twenty-foot great white got way too curious about our dive boat."

During the previous twenty years or so, every member of the team had done some research on *U-550* or had been on previous expeditions that tried to find the submarine. Mazraani said:

We got serious about finding the *U-550* when I bought *Tenacious*. In 2011 we made one search trip and did not locate it. We imaged the sub on sonar in 2012, almost toward the end of the expedition. We were pretty much wrapping up, having covered our planned search grid. Eric Takajian was driving the boat, and

The team that discovered *U-550*: from left to right, Joe Mazraani, Eric Takajian, Tom Packer, Steve Gatto, Anthony Tedeschi, Garry Kozak, and Bradley Sheard. The dive boat *Tenacious* is in the background. *via Bradley Sheard*

we had run twenty-four-seven for the past two days. We had finished our grid and decided to go back and investigate the large target that we passed over the night before. We were doing sonar passes over it, and that's when the boat's sonar picked it up. The submarine showed up on the boat's sonar before it showed up on the towfish, which is towed a distance behind the boat. By the time the towfish caught up to it, we had an idea that there was something there. It was just unbelievable.

We got some more images of it with sonar, then we used a drop camera to confirm that it was a submarine. The drop camera passed over the aft torpedo-loading hatch, which confirmed that the wreck was a Type IX U-boat. We went into port, then came out a week later and dove it, confirming it was the *U-550*.

The U-boat is located 330 feet below the surface, which is very close to the limits of technical diving. Recreational scuba breathing air has a limit of about 130 feet, so *U-550* requires extensive training to safely dive the wreck.

Some of us have rebreathers, and some of us are still diving with open-circuit equipment. We use several mixes. We include helium in some of the mixes because nitrogen at depth has a narcotic effect. Some helium is substituted as a lighter gas and doesn't have a narcotic effect. It allows you to function and think clearly. In a lot of mixes we include quite a bit of helium for those types of depths. We have to reduce the percentage of oxygen in our bloodstreams, because at a certain depth, oxygen can be toxic. We carry several bottles, two on our back and two to three on our sides, in order to take the decompression as we go up.

For this depth, we breathe a topside mix for the first twenty to thirty feet before switching to the bottom mix. The percentage of oxygen in our bottom mix for this particular dive is twelve percent, but that will not sustain life on the surface. You need at least seventeen percent oxygen on the surface. But once you get in the water, because of the pressure of the atmospheres and the water, as you start going deeper, that twelve percent gas increases to a sustainable breathing level. When you're on the boat walking to the end to jump off, you have to be breathing something that sustains life on the surface. Then, at about twenty feet down, you switch to your bottom mix. Then on the way up, you use whatever mixes you have worked out to do your decompression.

It takes about five minutes to descend the 330 feet to the wreck. If there's a current running, it may take longer, and divers with propulsion vehicles ("scooters") can get down in about three minutes. However, coming back to the surface is another matter. If a diver is using open-circuit scuba—in which they breathe in from the tanks and exhale the gas into the water—they are limited to fifteen, possibly twenty minutes of bottom time exploring the wreck. Mazraani said, "With rebreathers, we can get twenty-five to thirty minutes at three hundred and thirty feet, but the decompression penalty is three and one-half to four hours." Depending upon the diver's decompression program, they'll stop at two hundred feet and begin a slow ascent in ten-foot increments over the course of three and a half hours.

Bradley Sheard described his dive to the wreck by saying: "*U-550* is spectacular, but a bit scary as it is very deep. I had dived *U-550*'s only victim, *Pan Pennsylvania*,

back in 1994 and 1995—bow section of wreck—as well as being onboard the dive boat *Seeker* at that time when we looked for *U-550*, but unsuccessfully. So finding the sub and diving it was the fulfillment of a longstanding dream. We have returned the past two summers, and the dives have only gotten better with greater visibility and constantly improving camera technology, and we're now diving on closed-circuit rebreathers."

It would be quite dangerous to penetrate the wreck; thus, the team has used only a GoPro on a pole to film the inside of the U-boat. "It's just a jumbled mess in there—ready ammunition boxes, the bunk frames, and everything else scattered around from the thrashing the boat took from the depth charge attacks," Mazraani said.

U-550'S CONDITION AND HISTORICAL ASSESSMENT

"The conning tower is intact," said Mazraani. "However, the fairing around the port side of the conning tower is falling off. Both the navigation and the attack periscopes are there as well. The deck looks spectacular for having been down seventy years. I've never seen anything like it, a wreck so intact. It's unbelievable."

Mazraani reports that about fifteen feet of *U-550*'s stern is missing, confirming that the destroyer escort *Gandy* did, in fact, sever that section when it rammed the submarine. Lookouts on *Gandy* and *Peterson* both reported that torpedoes had been fired at them; however, Bradley Sheard lays that claim to rest: "Referencing Allied interrogation reports of the German survivors, as well as Hänert's report to his own navy, the crew was far too busy trying to save their badly leaking boat. There simply was no time or opportunity to fire a torpedo at the escorts."

The submarine's forward torpedo room, galley, and conning tower hatches are open. That is interesting because the Eastern Sea Frontier War Diary reports that the bodies of three of the U-boat men were found in the days following the sinking, some wearing submarine escape gear. "There's no way anyone escaped from that boat," Mazraani said. "There's a big crack on the forward hatch and that boat was not watertight when it sank. We went to Germany and interviewed two survivors. The chief engineer's still alive. We spoke to him, and he indicated that it was his duty to make sure the boat went to the bottom. He was the last person out of the sub. He checked every compartment before he gave it the last horn blast, so he can get out of there; there was nobody left in the boat. Only one person or maybe two people died in the gunfire up on the conning tower. Everybody else was alive. The destroyers only picked up thirteen. One died, the other twelve survived. The other members of the crew basically floated away and died at sea. They weren't picked up; they were left there. That's what they found, not survivors or escapees."

In his own estimation, Mazraani does not believe the written reports that after the destroyers ceased fire, the men on the U-boat began firing. "The Germans never had the opportunity to fight back or fire back. I know the reports say that they went to their guns. That's just not the case. When those sailors came out of the hatch, three destroyers surrounded them. I think what the destroyers felt was ricocheting

Diver Steve Gatto hovers over *U-550*'s conning tower, illuminating the intact RDF (radio direction finding) loop antenna. *Bradley Sheard* from ammo fired from the other escorts because they were surrounding the sub." Adding to Mazraani's body of evidence, he reports that the deck guns are still in the stowed position for cruising, not limbered up to shell the destroyers.

NEXT ON THE AGENDA?

What do divers do once they've found the most elusive of U-boat wrecks? Look for another missing ship, of course. "I'm particularly interested in finding a wreck called *Maiden Creek*. It foundered in a storm off of Long Island, New York, in 1942," Mazraani said.

Maiden Creek was a 5,500-ton riveted-steel-hull freighter built shortly after World War I at Hog Island, Pennsylvania, by American International Shipbuilding Corp. The freighter was 390 feet long with a beam of 54 feet and a draft of 27 feet. The vessel had five holds and could carry more than 365,000 cubic feet of cargo.

The freighter was carrying a cargo of copper and lead concentrate from Botwood Harbor, Newfoundland, Canada, to New York when she foundered on December 31, 1942. The approximate last position of the ship is given as fifty-eight miles south of Montauk, New York; however, *Maiden Creek* does not lie there. Fishing nets have caught on two underwater obstructions within a five-mile radius, each more than 220 feet below the surface, and both need to be investigated.

To wreck divers like Mazraani, finding *Maiden Creek* is another mystery to be solved. To researchers and scholars, the location and condition of the wreck are invaluable and another building block in the historical record.

The bow of *U-550* stands up straight on the bottom as a diver inspects the forward, portside torpedo hatches. *Bradley Sheard*

HUNTING
TROPHIES
U-BOATS IN CAPTIVITY TODAY

GERMANY'S *UNTERSEEBOOT*, OR U-BOAT FOR SHORT,
was the dominant weapon in the first half of the Battle of the Atlantic. From 1939 to 1943, U-boats ranged far and wide, sinking thousands of tons of ships and sending great amounts of war materiel to the bottom of the sea. Ranging the Atlantic Ocean with impunity during the war's early years, the brazen U-boats made many attacks so close to the US East Coast that American citizens could witness the sea battle with their own eyes. Oil-covered shipwreck survivors were washing up on the beach, yet nothing was being done to prevent the attacks.

The US Navy's Fleet Admiral Ernest J. King—who simultaneously served as commander-in-chief, United States Fleet, and as chief of naval operations (appointed on March 18, 1942)—disliked the British and everything about them. It was King who delayed blacking out East Coast cities, which would have reduced the number of sinkings by preventing ships from being backlit as they traveled between the shore and lurking U-boats in deeper waters. Even if the ship was steaming blacked out, all a U-boat had to do was wait until the city lights were blocked to pick up a target. In addition, Admiral King delayed sailing merchant ships in convoys until May 1942, when he impressed a number of small ships and other vessels to serve as escorts. While King was contemplating how to protect merchant shipping, US shipyards began to turn out destroyers and destroyer escorts that would form the backbone of convoy escort forces. Illustrating the point that the Allies were ill prepared for and overwhelmed by the U-boat threat in 1942, German submarines sent 1,859 ships—or 8.338 million tons of shipping—to the bottom that year.

The tide in the Battle of the Atlantic began to change in 1943, the first year the Allies built more ships than the Nazis sank. In the final tally, from 1939 to 1945, the Nazis commissioned 1,171 U-boats. They lost 761 U-boats to all causes, and 156 were surrendered at the end of the war. The German navy's Operation *Regenbogen* saw 195 U-boats scuttled in the Baltic Sea by their crews in advance of the

May 5, 1945, Nazi surrender. After the war, 116 were sent to the bottom in the Allies' Operation Deadlight, and 12 were sunk in postwar weapons tests by the victors.

A small number were put into service by Allied navies. The British Royal Navy tested a number of U-boats, including *U-792*, *U-793*, *U-953*, *U-1054*, *U-1105* (as *N-16* in Royal Navy service), *U-1108*, *U-1171* (*N-19*), *U-1407* (HMS *N-25* and later HMS *Meteorite*), *U-2326* (*N-35*), and *U-3017* (*N-41*). All were scrapped by the end of 1949.

The German Federal Navy raised two Type XXIII U-boats—*U-2365* was brought up in June 1956, and became *U-Hai* (*S-170*). The boat sank again on September 14, 1966, taking nineteen men with her. She was raised five days later and then scrapped. It also raised *U-2367*, in August 1956. It was refurbished, named *U-Hecht*, and served until September 1968.

The French acquired five U-boats that they sailed in the postwar years. *U-510*, a Type IXC surrendered at St. Nazaire on May 9, 1945, was overhauled and operated with the French navy as *Bouan* (later *Q176*) until May 1, 1959. The submarine was scrapped one year later. A second U-boat, Type XXI *U-2513*, entered service with the French navy on February 14, 1946, as *Roland Morillot* (later *Q426*). She was phased out of service in late 1967 and was scrapped two years later. *U-766* was surrendered to France and after overhaul became *Laubie* (later *Q335*), serving until 1963, when she was scrapped. *U-471* was heavily damaged in an air raid on Toulon, and it, too, was surrendered to the French navy. This boat entered service with the French in 1946 as *Millé* (later *Q339*) and served until 1963. The French also acquired the Type IXB *U-123*, which played an important part in Operation *Paukenschlag* (Drumbeat)—shipping attacks off the US East Coast. It made twelve war patrols, sinking forty-two ships (219,924 gross registered tons), and was turned over to the French at Lorient. After overhaul, this boat became *Blaison* (later *Q165*) and served until 1959. This historic submarine was cut up the following year.

U-995 is credited with sinking four ships, all in Arctic waters. At the end of the war, she was surrendered to the British, who in turn gave the submarine to the Norwegian navy. The Norwegians overhauled the boat and placed her into service as the *Kaura*. In 1965, they returned the submarine to Germany, where she is now one of the featured artifacts at the German Naval Memorial at Laboe. *Deutscher Marinebund e.V.*

The Spanish navy acquired Type VIIC *U-573*, which was heavily damaged by British aircraft on May 1, 1942. Unable to make the safety of a German-controlled port on the French coast, *U-573* limped into Cartagena, Spain, the following day. The Spanish interned the sub, and subsequently acquired it from Germany on August 2, 1942. The Spanish designated the former U-boat "G 7" and sailed her for the next twenty-eight years. It, too, was scrapped.

U-926, a Type VIIC, was surrendered to the Norwegian navy on May 9, 1945. Refitted, the sub was renamed KNM *Kya* (KNM standing for *Kongelige Norske Marine*, the Royal Norwegian Navy) and served until 1962. Another Type VIIC, *U-1202*, was renamed KNM *Kinn*, entering service on July 1, 1951. *Kinn* served for a decade and was transferred to Hamburg for a potential memorial, but was scrapped in 1963. Type XXIII *U-4706* became KNM *Knerter*. Her service was short-lived, and she was scrapped in 1955.

The Soviet Union operated the Type VIIC *U-1057* as *S-81*. The sub sailed until September 1955, when the Soviets used it as an atomic test target. In a subsequent atomic blast in September 1957, the sub was heavily damaged and was soon scrapped. *U-1058* was also allocated to the Soviets after the war. This sub became *S-82* and served until January 1956. *S-83*, the former Type VIIC *U-1064*, served until March 1974. *U-1305*, a Type VIIC/41, became *S-84* and was sunk in the same October 1957 tests in the Barents Sea.

Four Type XXIs and a single Type XXIII were also operated by the Soviets. *U-2529*, a Type XXI, served with the Baltic Fleet, designated *BSh-28* and later *UTS-3*, until fall 1972. The other Type XXIs operated by the Soviets included *U-3035*, *U-3041*, and *U-3515*. *U-2353*, a Type XXIII, was transferred to the Soviets on December 4, 1945. The sub served with the Baltic Fleet, and in June 1949 its name was changed to *M-51*. The obsolete boat was scrapped in 1963.

U-BOATS IN CAPTIVITY

By 1975, nearly all of the U-boats that survived to see postwar service with the British, French, German, Norwegian, Spanish, and Soviet navies had been cut up for scrap. Today, only four U-boats survive on display along with twenty-eight Nazi midget submarines of four different types.

U-995 is the only surviving Type VIIC (Type VIIC/41) U-boat. The submarine was built by Blohm und Voss Shipyards in Hamburg and was commissioned on September 2, 1943. While under construction, the Blohm und Voss yard was bombed, damaging the unfinished U-boat. On May 21, 1944, *U-995* was sailing north to Trondheim, Norway, to begin operations in Arctic waters. The sub was strafed by a Short Sunderland flying boat from Operational Training Unit Squadron 4, Royal Canadian Air Force. The plane's gunners wounded four of the sub's crew and did some damage to the boat. *U-995* spent its entire career in the Arctic and is credited with sinking two large freighters (each approximately 6,000 tons), one 2,000-ton freighter, and one Soviet navy patrol craft (105 tons) and damaging the American Liberty ship *Horace Bushnell*. *U-995* should have been given full credit for this attack as *Horace Bushnell* was beached near Russia's Kola Inlet and spent the rest of the war there.

On May 9, 1945, *U-995* was surrendered to the British Royal Navy at Trondheim, Norway. The sub had been opened up to install a snorkel, and the British transferred

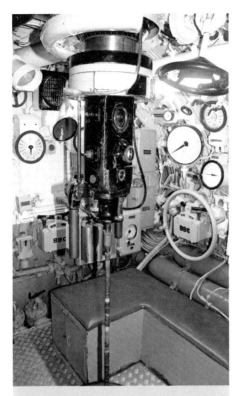

The "lookout," or surface-search periscope, in the control room of *U-995*. This periscope has a wider field of view than the attack periscope located in the conning tower one deck above. *Nicholas A. Veronico*

the unseaworthy submarine to the Norwegian navy, which placed the boat into storage. The sub was overhauled and placed into Norwegian navy service as KNM *Kaura* on December 1, 1952.

Her career ended in 1965, and she was subsequently returned to Germany. After modification back to Type VII/41 configuration, which took six years, *U-995* was placed on display at the Laboe Naval Memorial, Germany. The U-boat Memorial at Möltenort, Germany, which contains the names of more than thirty-five thousand German submariners from both world wars, is a few miles away.

U-2540: AN ADVANCEMENT IN SUBMARINE TECHNOLOGY

The Type XXI submarine was designed to be quieter and stay submerged longer than its contemporary U-boats. The sub's hull and conning tower featured big improvements in hydrodynamic stream-lining, which in turn not only made the boat faster underwater but reduced drag, thereby increasing range. Type XXI *U-2540* had a brief career with the German *Kriegsmarine*, having been commissioned on February 24, 1945. Fuel shortages in the closing months of the war kept the boat in home waters, and she was scuttled near the mouth of the Flensburg Firth, which divides Germany from Denmark, on May 4, 1945.

In 1957, *U-2540* was raised from the sea and overhauled at Kiel. The submarine reentered service in 1960 and was used as a research vessel until 1968, when she was overhauled again and renamed *Wilhelm Bauer*. It served until 1980, and was listed for sale in 1982. The German Maritime Museum at Bremerhaven acquired *U-2540* and restored her back to its wartime configuration. The boat was placed on display on April 27, 1984.

U-505: FIRST FOREIGN MAN-O'-WAR CAPTURED SINCE 1815

In November 1943, *Oberleutnant zur See** Harald Lange assumed command of *U-505*, a Type IXC built at Deutsche Werft in Hamburg. It had been deemed a "hard luck" submarine, and Lange was sent to the boat to change things.

The boat had been commissioned on August 26, 1941, with Axel-Olaf Loewe in command. Loewe was an aggressive captain, and he and his crew were responsible

---

* *Oberleutnant zur See (Oblt.z.S): equivalent to a US Navy lieutenant, junior grade*

SPECIFICATIONS: PRESERVED U-BOATS

	U-505	U-534	U-995	U-2540
Type	IXC	IXC/40	VIIC/41	XXI
Length, overall	251 feet 10 inches	251 feet 10 inches	220 feet 0 inches	251 feet 8 inches
Beam	22 feet 2 inches	22 feet 2 inches	20 feet 4 inches	26 feet 3 inches
Draft	15 feet 5 inches	15 feet 5 inches	15 feet 7 inches	17 feet 5 inches
Displacement, sufaced/ submerged	1,120 tons/ 1,232 tons	1,120 tons/ 1,232 tons	747 tons/ 846 tons	1,621 tons/ 2,100 tons
Powerplant	2 MAN M9V 40/46 supercharged 9-cylinder diesel engines of 4,000 hp; 2 electric motors capable of 1,000 hp submerged	2 MAN M9V 40/46 supercharged 9-cylinder diesel engines of 4,000 hp; 2 electric motors capable of 1,000 hp submerged	2 supercharged Germaniawerft 6-cylinder diesel engines of 2,800 hp; 2 electric motors capable of 750 shaft horsepower at 296 rpm	2 MAN M6V40 supercharged 6-cylinder diesel engines of 4,000 hp; 2 SSW GU365/30 electric motors (3.7 megawatts); and 2 SSW GV232/28 electric motors (0.166 megawatts)
Speed, surfaced/ submerged	18.3 knots/ 7.3 knots	19 knots/ 7.3 knots	17.7 knots/ 7.6 knots	15.6 knots/ 17.2 knots
Range	13,450 miles at 10 knots	13,850 miles at 10 knots	8,190 miles at 10 knots	17,800 miles at 10 knots
Maximun depth	755 feet	755 feet	820 feet	919 feet
Crew	56 officers and sailors	56 officers and sailors	52 officers and sailors	60 officers and sailors
Armament	105mm deck gun 4 bow and 2 stern torpedo tubes 22 torpedoes	105mm deck gun 4 bow and 2 stern torpedo tubes 22 torpedoes	4 bow and 1 stern torpedo tubes 12 torpedoes or 39 mines 1 37mm and 2 20mm anti-aircraft guns	6 bow torpedo tubes 23 21-inch torpedoes 4 20mm anti-aircraft guns
Builder	Deutsche Werft AG, Hamburg	Deutsche Werft AG, Hamburg	Bloom & Voss, Hamburg	Bloom & Voss, Hamburg
Launched	May 24, 1941	September 23, 1942	July 22, 1943	January 13, 1945
Commissioned	August 26, 1941	December 23, 1942	September 16, 1943	February 24, 1945
Fate	Captured June 4, 1944	Sunk May 5, 1945	Surrendered May 9, 1945	Scuttled May 4, 1945
Preserved by	Museum of Science and Industry, Chicago, Illinois	U-Boat Story, Woodside, England	Laboe Naval Memorial, Laboe, Germany	Technikmuseum U-Boat Wilhelm Bauer, Bremerhaven, Germany
Website	msichicago.org	u-boatstory.co.uk	deutscher-marinebund.de/ u995_geschichte_ english.htm	bremerhaven.de/en or u-boot-wilhelm-bauer.de
Coordinates	Inside the museum building	59.395143, -3.009371	54.412484, 10.228956	53.541380, 8.577774

The Type XXI was Germany's most advanced submarine of the war and it was faster submerged than surfaced. Speed underwater often meant survival. This submarine, *U-2540*, was scuttled on May 4, 1945. Twelve years later, she was raised and refurbished, entering service in 1960 as the research submarine *Wilhelm Bauer*. *U-2540* has been restored back to her wartime configuration and is now displayed at the German Maritime Museum in Bremerhaven. *Nicholas A. Veronico*

for the destruction of seven ships (totaling 37,789 tons) by the end of July 1942. Loewe suffered an appendicitis attack in August 1942, cutting the boat's patrol short.

Kapitänleutnant Peter Zschech assumed command in September 1942 and took the boat out from Lorient, France, to patrol in the Caribbean Sea north of the island of Trinidad. Zschech sent the 7,173-ton freighter *Ocean Justice* to the bottom east of Trinidad on November 7, 1942. Three days later, the boat was surprised on the surface by a Royal Air Force Lockheed Hudson twin-engine patrol bomber. The Hudson made a direct hit with a 250-pound bomb that damaged the deck and pressure hull, destroying the antiaircraft gun and killing one and wounding another on deck. With the boat taking on water, Zschech ordered the crew to abandon ship; however, they attempted to save the boat, and after two weeks were able to cut away the damaged sections and made the submarine watertight. The boat limped back to Lorient for repairs that eventually took six months.

U-505's next four patrols were all cut short after only a few days at sea. The boat was forced to make the dangerous transit back to Lorient across the Bay of Biscay, the air above which was owned by the Allies. Many of the problems were ascribed to the French resistance, whose members had infiltrated the docks and repair yards in the area.

Zschech and the yard crew finally got all the boat issues squared away, and *U-505* departed on its tenth war patrol on October 9, 1943. The submarine was again caught on the surface, this time on October 24, 1943. Steaming near the Azores, Royal Navy destroyers pummeled *U-505* with depth charges. Zschech, unable to bear the strain of the attack, committed suicide in the control room—in full view of his crew. The first watch officer took command of the boat, weathered the depth-charge attack, and returned the aubmarine to Lorient.

Having logged heavy damage by patrolling aircraft, multiple equipment failures blamed on French partisans, and a commander who did not handle the stress of war, it was time for a new commander, and Harald Lange was that man.

As Lange sailed *U-505* on its final war patrol, Allied code breakers were monitoring the deployment of German submarines. The Allies had determined that certain submarines were operating off the west coast of Africa, and they sent Capt. Daniel V. Gallery's Task Group 22.3 in pursuit of the U-boats. Task Group 22.3 was composed around the escort carrier *Guadalcanal* (CVE-60) and Escort Destroyer Division 4 consisting of *Pillsbury* (DE-133), *Chatelain* (DE-149), *Flaherty* (DE-135), *Jenks* (DE-665), and *Pope* (DE-134).

On a previous patrol, in March 1943, Gallery's task group had sunk *U-515* and *U-68*. During an after-action review, it was determined that *U-515* potentially could have been boarded and its sinking prevented—and all of its codes and technologies captured by the Allies.

In a May 26, 1945, narrative of the capture of *U-505*, Gallery detailed that a U-boat's capture had been set in motion before Task Group 22.3 left the dock:

> The next cruise started in May and at the departure conference attended by the representatives of Cinclant [Commander in Chief, US Atlantic Fleet], ComAirLant [Commander, Atlantic Air Forces], ComFairNorfolk [Commander, Fleet Air, Norfolk, Virginia], DesDivLant [Destroyer Division, Atlantic], we discussed plans for boarding and capture and we agreed that if we encountered a submerged submarine and forced him to surface we would then assume that he had surfaced for only one reason, which was to try to save his hide, save the crew.
>
> As soon as [the U-boat] surfaced we would cease fire with any weapons that could inflict fatal damage on the submarine, and we would use only anti-personnel

Destroyer escort *Pillsbury* (DE-133) comes alongside *U-505*, which has just been captured by Capt. Daniel V. Gallery's Task Group 22.3 off the west coast of Africa. The captured U-boat was towed to Bermuda, where it could be hidden from prying eyes to keep the Germans from knowing it had fallen into American hands. *US Navy*

U-505 is seen as a towline is rigged shortly after capture. With its deck awash, its future was still in doubt. Lieutenant Junior Grade Albert L. David from *Pillsbury* led a group of sailors into the sinking submarine to recover its Enigma code machine and code books. *U-505* was the first enemy ship captured since June 1815. *US Navy*

weapons from that point on, attempting to drive the crew overboard as rapidly as possible, meantime having our boarding parties already to go. Each ship in the task force was ordered to organize and instruct boarding parties and to have all preparations ready for taking a sub in tow on short notice. On May 17, approximately ten days after we had sailed, in my intention for the night signal I told the group that we expected to be on a hot trail the next day and reminded them that our objective was to capture rather than sink, and said for all ships to have their boarding parties ready and be ready to tow.

Gallery's Task Group 22.3 patrolled the waters around the Cape Verde Islands in search of the U-boats reported to be in the area. Unable to locate the marauding German sub, the task group turned toward Casablanca, Morocco, to refuel. Task Group 22.3 searched for four days as it steamed toward its intended refueling point.

On the morning of June 4, 1943, at 11:10 a.m., approximately 150 miles west of Cape Blanco, French West Africa (at the southern border of Morocco), *Chatelain* detected a contact that was quickly identified as a submarine. *Guadalcanal* directed its fighters to the area around *Chatelain*. *Pillsbury* and *Jenks* assisted *Chatelain* while *Guadalcanal* and its two escorting destroyers (*Flaherty* and *Pope*) turned to launch additional aircraft. At 11:16, *Chatelain* ran down the contact and fired a pattern of Hedgehog depth bombs, none of which impacted the target submarine.

The battle developed very quickly. The following timeline was extracted from the war diaries of Task Group 22.3 and destroyer escort *Chatelain*:

11:21 a.m.:   "Fired fourteen-charge shallow pattern depth charges."

11:22.5 a.m.:   "U-boat surfaced, bearing 083°True, 8,000 yards from the carrier. Two VFs [Wildcat fighters] strafed and escorts opened fire."

11:23 a.m.:   Radio transmission from Commander, Task Group 22.3—Capt. Daniel V. Gallery—to ComCortDiv 4 [Commander, Escort Division Four]: "I would like to capture that bastard if possible."

11:24 a.m.:   Escort Carrier *Guadalcanal*: "Launched killer group of one VT [Avenger torpedo bomber] and 1 VF [Wildcat fighter]."

11:25 a.m.:   Destroyer Escort: "*Chatelain* reported U-boat hit several times. *Jenks* ordered to cease fire."

11:27 a.m.:   "*Chatelain* reported, 'They are all holding their hands up. They are surrendering.'"

11:38 a.m.:   "Escorts commenced recovering survivors from the water."

11:42 a.m.:   "Boarding party alongside U-boat."

That boarding party was led by Lt. j.g. Albert L. David, USN, assistant engineering officer from *Pillsbury*. They quickly secured the Enigma cypher machines and code books and other documents, then began the task of securing the boat. While trying to stem the on-rushing water, Motor Machinist Mate Zenon B. Lukosius found the cover to the sea strainer and replaced it. Lukosius's actions greatly reduced the flooding and bought the boarding crew more time to secure the remaining valves. Crews worked for hours to pump out the boat and get her to a more even keel.

Soon a party from *Guadalcanal* rigged a line to the sub and began towing her to keep her afloat while the pumps went to work. The original intention was to tow *U-505* to Casablanca. With too many prying eyes in the African port, it was decided to deliver *U-505* to the port at Bermuda, where she could be hidden from the Germans for the duration of the war.

U-505 was the first enemy combatant ship to be captured at sea by the US Navy since June 30, 1815, when the US Navy sloop *Peacock* boarded and captured the East India Company's *Nautilus*.

After the war, *U-505* was towed to the Portsmouth Navy Yard, New Hampshire, where she sat moored to a pier. Gallery, recently promoted to rear admiral, became aware of the fact that *U-505* was slated to be sunk as a torpedo target. Gallery told his brother, Father John Gallery, who in turn contacted the Museum of Science and Industry in Chicago to see if they could offer salvation to the historic former wartime submarine. After a number of years raising the funds to transport the sub from the Atlantic coast to Lake Michigan, the U-boat was loaned to the museum in 1954. She was restored and displayed outside from 1954 to 2004, when *U-505* was moved to a new, purpose-built exhibit resembling a U-boat pen. Visitors can tour the interior of the submarine and interact with educational displays outside the boat.

In September 2002, as plans were in the works to move *U-505*, its original aerial/navigation periscope was located at the US Navy's Space and Naval Warfare Systems Center, in San Diego, California. The periscope had been removed from *U-505* and studied by US Navy technicians at the San Diego facility. Once its secrets had been unlocked, the periscope was built into an underwater arctic testing laboratory. When the lab was to be demolished, experts at the center realized its heritage and made contact with the Museum of Science and Industry. The periscope was loaned to the museum by the Navy. Still missing is the boat's attack periscope, which curators hope will be located soon.

U-534 TELLS "THE U-BOAT STORY"

The Type IXC/40 *U-534* was ordered on April 10, 1941, and laid down the following year on February 20. Seven months and three days later, on September 23, the boat slid down the ways at Deutsche Werft AG, in Hamburg. The boat was commissioned on December 23 with *Oblt.z.S* Herbert Nollau as the commanding officer. Nollau was later promoted to the rank of *Kapitänleutnant*.

After the war, *U-505* was slated to serve as a target for an American submarine in a live-fire exercise. The former enemy submarine was saved and moved from Portsmouth, New Hampshire, to the Museum of Science and Industry in Chicago. *U-505* is seen here during the July 4, 1954, weekend as it is prepared for the move across Lake Shore Drive. *Author's collection*

Nollau had been executive officer of *U-505* from August 1941 to September 1942, and he was promoted from that boat to become the captain of *U-534*. He skippered *U-534* through its training period (December 23, 1942, to May 31, 1943) and its wartime service with 2 Flotille and 33 Flotille.

After six months of crew training, *U-534* was reconfigured by having its main deck gun removed and additional antiaircraft guns added. Three months later, on June 1, 1943, *U-534* was assigned to the 2nd Flotilla based at Lorient, France, and her first war patrol took the boat to the waters off Greenland to provide weather reports to the German armed forces. While sailing home to the submarine pens at Bordeaux, France, *U-534*, *U-437*, and *U-857* were attacked on the surface by a pair of Royal Air Force bombers. The bombers retired without inflicting heavy damage on the trio of subs.

Fitted with a new snorkel that allowed fresh air into the boat to drive the diesel engines while the submarine cruised submerged, *U-534* headed out to sea. When the snorkel was tested on October 25, 1944, exhaust fumes back-flowed into the boat and Nollau was forced to surface. A Wellington bomber (RAF serial NB798) from 172 Squadron attacked the sub but was shot down by the U-boat's gunners.

On May 1, 1945, *U-534* departed Kiel carrying three T-11 *Zaukönig* torpedoes that could home in on a specific sound; they were the most effective homing torpedoes of the war. Only thirty-eight of the T-11s are thought to have been built, and many believe *U-534*'s departure on May 1, 1945, was to deliver the secret torpedoes to the Japanese.

On May 5, after the surrender order had been given, *U-534* was cruising on the surface of the Kattegat Sea (between Denmark and Sweden), northeast of the island of Anholt, in the company of *U-3503* and *U-3523*. *U-534* was attacked by Royal Air Force Liberator bombers, one of which was shot down. The Liberators dropped nine depth charges that missed but hit with a tenth, dropped by Liberator G (George), exploding under the stern of *U-534*. The aft torpedo room was breached and began to flood, and the sub sank stern first. All of the crew escaped except for five men in the forward torpedo room. Once the submarine settled on the bottom, two hundred feet below the surface, the men equalized the pressure and escaped the submarine. One man died during the ascent, and two others later perished while awaiting rescue.

U-534 was sunk on May 5, 1945, after the surrender order had been broadcast to all U-boats at sea. After recovery, the sub ended up at the Birkenhead Docks as part of the Warship Preservation Trust. When that group's display closed, the submarine's fate was in doubt. *The U-Boat Story/Merseytravel*

The former German U-boat was acquired by Merseytravel, which intended to build an attraction around it at the Woodside ferry terminal, across the Mersey River from Liverpool. The first step was to strip the hull, apply rustproofing, then repaint it. *The U-Boat Story/Merseytravel*

Forty-one years after the sinking, in 1986, Danish wreck diver Aage Jensen discovered the wreck. With *U-534*'s departure days before the end of the war, many had speculated the sub was headed to South America with gold or stolen artwork. A $6.25 million (£4 million) salvage effort was funded by Danish businessman Karsten Ree and coordinated by Lars Sunn Pedersen. Since none of the crew perished in the hull, the ship was not considered a war grave and was raised from the sea floor by the Danish salvage company Smit Tak, which worked closely with the Danish navy.

Once the interior was inspected, the documents and artifacts were removed for conservation. Unfortunately, no treasure was found inside the submarine. In 1996, *U-534* was acquired by the Warship Preservation Trust and displayed at the Birkenhead Docks in England. A decade later, on March 1, 2006, Warship Preservation Trust's display closed and *U-534*'s fate hung in the balance.

Merseytravel, the Liverpool-area transit authority, rescued the U-boat with the intention of developing a

Moving the 235-foot-long submarine in one piece was cost prohibitive and the sub would not fit into the available display space, so the decision was made to saw it into five pieces. A diamond wire-cutting system was used to precisely cut the submarine forward or aft of certain bulkheads. The precision cuts allow for it to be easily reassembled in the future. *The U-Boat Story/Merseytravel*

Once cut into transportable pieces, each 230-ton section was individually loaded and transported by the floating crane "Mersey Mammoth." Only one section could be moved per day due to tides and river currents. *The U-Boat Story/Merseytravel*

display around the former warship. The authority's aim was to retain the submarine, restore it, relocate it, and provide a unique visitor attraction for the area. The space chosen for the display, known as "The U-Boat Story," was the ferry terminal at Woodside, on the River Mersey, opposite the city of Liverpool.

The first step was to restore the exterior of the submarine to prevent further deterioration. Rustproofing was applied to the hull, which was then primed and painted

U-534 is now displayed at the Woodside Ferry Terminal and is known as "The U-Boat Story." Each section has been capped with glass and the interior illuminated, enabling guests to see inside the submarine. Although it was cut into five sections, the after torpedo room and stern machinery spaces have been reassembled, presenting the submarine in four sections. An indoor, interactive museum showcases the restored artifacts found inside the submarine. *The U-Boat Story/Merseytravel*

gray. Merseytravel decided that transporting the 235-foot-long submarine in one piece would be difficult and cost prohibitive. The authority chose to cut the submarine into five sections and, once at its new location, display it as four distinct pieces.

Marine engineers for the project were Royal HaskoningDHV of Amersfoort, Netherlands, and Holemasters Demtech Ltd. cut the hull. A diamond wire-cutting system was used to make ten-millimeter cuts. The cutting system uses a steel cable wrapped in coiled wire coated in rubber that is embedded with diamonds and turned through a series of pulleys. The cuts, which took five days, preserved the integrity of the vessel and allowed for the sub to be reassembled back into a single-hull configuration if that was desired at a later time.

Each of the sections weighed 230 tons, and they were lifted and transported to Woodside one section at a time using the "Mersey Mammoth" floating crane. The engineers had to plan the lifts taking the river tides and currents into consideration. Only one section could be lifted, moved, and deposited per day, and the operation took six days as one was lost to bad weather. Once the U-boat's sections were delivered to Woodside, they were lowered onto self-propelled modular transporters for the trip into position. Wheels of the transports can rotate through 90° to enable exact positioning of the hull sections. The hull pieces were then lowered using hydraulic jacks.

When the submarine's sections were positioned on the docks, the aft torpedo room and the stern machinery spaces were joined together. Now in four sections, each opening has been sealed with glass panels and illuminated from the inside, giving the public an excellent view of the fighting submarine's equipment and cramped spaces. Visitors can also see the lower sections of the hull showing the battle damage on the starboard stern section and the outer torpedo tube doors at the bow.

One of the once-top-secret T-11 torpedoes recovered from *U-534* is on display in the outdoor gallery and a new, $7.8 million (£5 million) interactive interior display gallery shows movies and interviews with surviving *U-534* crewmembers and the aircrew of Liberator G (George) that sank the U-boat. Memorabilia on display includes personal artifacts, an Enigma machine, medals, dishes, and documents recovered from inside the submarine.

Now that *U-534*'s future is secure at the Woodside Ferry Terminal, she joins the short list of preserved and displayed U-boats that represent the sacrifices made by Axis and Allied sailors during World War II.

The U-Boat Story has a promenade at deck level and guests can descend a staircase to see the submarine's lower exterior. The propellers, rudder, and diving planes can be examined up close from this vantage point. *The U-Boat Story/Merseytravel*

ONCE WITHIN STRIKING
DISTANCE OF TRUK ATOLL,
A MASSIVE FIGHTER ATTACK
WAS LAUNCHED TO CLEAR THE
SKIES AND DESTROY AS MANY
AIRCRAFT ON THE GROUND
AS POSSIBLE. WHILE OTHER
FIGHTERS FLEW COMBAT AIR
PATROL OVER THE FLEET, ADDI-
TIONAL FIGHTERS PROVIDED
TOP COVER FOR THE DIVE- AND
TORPEDO-BOMBERS THAT WENT
IN NEXT TO ATTACK SHIPPING
IN THE ANCHORAGE.

SINKING THE
BIG FLEETS

SINK EVERYTHING IN SIGHT

OPERATION HAILSTONE

OPERATION HAILSTONE WAS THE CODE NAME FOR the American attack on the Japanese naval base at Truk Atoll in the Caroline Islands. Aircraft from US Navy fast carrier task forces inflicted heavy damage upon the enemy fleet anchored in the lagoon, and for the Japanese, it was a bloodbath.

During World War I, the Japanese seized German colonies in Micronesia, including the Caroline, Marianas, Marshall, and Palau island groups. At the end of the war, the League of Nations Class C Mandate, known as the South Pacific Mandate, saw the former German possessions transferred to Japan.

Beginning in the 1930s, the Japanese fortified Truk Atoll and the entrances to its 35-mile-diameter (105-mile-circumference) lagoon. There are six entrances to the lagoon that lead sailors to the southwestern islands, known as the Shiki group, whose major islands (largest to smallest) are Moen, Dublon, Eten, Fefan, Uman, and Param; or to the east and southeastern islands, the Shichiyo group, including Udot, Ulalu, and Tol. In total, twenty-five islands make up the two groups, most of which are ringed by wide coral reefs with mangrove swamps that lead to hilly terrain. The atoll's islands have three peaks taller than 1,100 feet—Mount Tumuital on Tol Island (1,483 feet), Mount Teroken on the central part of Moen Island (1,214 feet), and Mount Tolomen on the southwest part of Dublon Island (1,184 feet)—plus another half dozen that reach heights between 700 and 1,000 feet. The northern part of the atoll is ringed by a series of small, uninhabited coral islets. All of these geographic features combine to make Truk Atoll one of the best natural anchorages in all of the Pacific Ocean.

Truk Lagoon, today known as Chuuk Lagoon, was the Japanese navy's major supply center and anchorage for the central and south Pacific Ocean. Extensive construction had been carried out on the islands in the prewar years, including a variety of warehouses and bunkers, ship repair facilities, a radar station, and five airfields. The Guadalcanal and New Guinea invasions were supplied from Truk, as well as many smaller island garrisons. Until the beginning of 1944, the Japanese felt the

124

atoll was impervious to attack due to the more than 350 aircraft stationed there as well as its distance from most Allied-held island bases. Knowing that it was so heavily fortified earned it the Allied nickname "the Gibraltar of the Pacific."

Because of its remote location, little was known about Truk Lagoon and its facilities until early 1944 when the Allies took Bougainville in the Solomon Islands. A base was constructed on Stirling Island, putting long-range B-24/PB4Y-1 Liberators within one thousand miles (each way) of Truk. The first Navy PB4Y-1 reconnaissance flight over Truk Lagoon took place on February 4, 1944, when a pair of planes from Marine Photoreconnaissance Squadron 254 (VMD-254) overflew the atoll, bringing back photos of a strong Japanese force at anchor—including photos of the battleship *Musahi*—and the atoll's shore installations.

Task Force 58, under the command of Vice Adm. Raymond A. Spruance, was assigned to raid the atoll, the attack taking place on February 16 and 17, 1944 (US dates). Task Force 58 was built around five fleet aircraft carriers, three light carriers, six battleships, three cruisers, and twenty-seven destroyers, divided into three task groups. Rear Admiral J. V. Reeves commanded Task Group 58.1, composed of carriers *Enterprise* (CV-6), *Yorktown* (CV-10), *Belleau Wood* (CVL-24); four cruisers; and nine destroyers. Rear Admiral Albert E. Montgomery's Task Group 58.2 comprised carriers *Essex* (CV-9), *Intrepid* (CV-11), and *Cabot* (CVL-28) plus four cruisers and seven destroyers. And Rear Adm. Frederic C. Sherman's Task Group 58.3 was carriers *Bunker Hill* (CVL-17), *Cowpens* (CVL-25), and *Monterey* (CVL-26); battleships *Alabama* (BB-60), *Iowa* (BB-61), *Massachusetts* (BB-59), *New Jersey* (BB-62), *North Carolina* (BB-55), and *South Dakota* (BB-57); two cruisers; and eleven destroyers. The Task Force sortied from Kwajalein on February 11, then refueled three days later en route to Truk. It was supported by seven submarines from Task Force 17 and three from Task Force 72 that patrolled the waters around the atoll.

Concerned by the reconnaissance flight over the lagoon, American submarine activity in the waters outside the lagoon, and Japanese radio direction finding tracking of US ship movements, the Japanese Imperial Navy quickly moved its warships out of Truk Lagoon in the days before the attack. Most of the merchant ships were held in the harbor due to high seas. The departure of the warships, however, didn't stop the onslaught of American ships and aircraft descending on the atoll.

A couple of warships that left the atoll late paid the ultimate price. During the morning of February 16, the carriers *Bunker Hill* and *Cowpens* dispatched SB2C Helldivers from VB-17 and TBF Avengers from VT-17 and VT-25 to sink the Japanese light carrier *Naka* in open water, thirty-five nautical miles west of Truk. The training cruiser *Katori*, destroyer *Maikaze*, submarine chaser *Ch-24*, and armed merchant cruiser *Akagi Maru* were all caught in open water outside the atoll and sunk that morning as well.

Once within striking distance of Truk Atoll, a massive fighter attack was launched to clear the skies and destroy as many aircraft on the ground as possible. While other fighters flew combat air patrol over the fleet, additional fighters provided top cover for the dive- and torpedo-bombers that went in next to attack shipping in the anchorage. At the end of the day, the following Japanese ships lay on the bottom of Truk Lagoon: submarine depot repair ship *Heian Maru*; aircraft transport *Fujikawa*

Another Japanese ship is destroyed in the Truk Lagoon anchorage, east of Dublon Island, by planes from the fast carriers of Task Force 58. In addition to the direct hit that is engulfing nearly the entire ship, three near-misses can be counted in the water at the stern and one bomb has just exploded, sending a water column into the air. *US Navy*

Strike photo from an *Essex* TBF Avenger bomber with a small cargo ship disappearing behind the smoke and sea spray of a direct hit. Mount Tolomen on Dublon Island dominates the terrain in the background. *US Navy*

Maru; transports *Aikoku Maru*, *Amagisan Maru*, *Gosei Maru*, *Hanakawa Maru*, *Hokuyō Maru*, *Kenshō Maru*, *Kiyozumi Maru*, *Matsutani Maru*, *Momokawa Maru*, *Reiyo Maru*, *Rio de Janeiro Maru*, *San Francisco Maru*, *Seikō Maru*, *Taihō Maru*, *Yamagiri Maru*; *Zuikai Maru*, and *No. 6 Unkai Maru*; fleet tankers *Fujisan Maru*, *Hoyo Maru*, *Shinkoku Maru* (10,020 tons, notable for having supported Admiral Nagumo's Pearl Harbor strike force), and *No. 3 Tonan Maru*; water carrier *Nippō Maru*; auxiliary *Yamakisan Maru*; Japanese army transports *Nagano Maru* and *Yubai Maru*; and freighter *Taikichi Maru*. In addition to the tonnage sent to the bottom, the Japanese lost a large number of experienced sailors during the attack—men not easily replaced.

Early the following morning, a single Japanese bomber was able to get close enough to the carrier *Intrepid* to launch an aerial torpedo. The underwater missile struck the carrier on the starboard side, near the rear of the ship. The torpedo's explosion killed eleven men instantly, caused flooding, and damaged the carrier's rudder. The flooding was sealed off, but the rudder could not be repaired at sea and *Intrepid* was sent to Pearl Harbor for repairs.

Later that afternoon, the second day of attacks, planes from Task Force 58 sent the destroyer *Fumizuki*, submarine chaser *Ch-29*, and motor torpedo boat (*Gyoraitei*) No. 10 to the bottom of the lagoon.

More than two hundred thousand tons of Japanese shipping, both warships and cargo ships, were sent to the bottom during the two-day raid. In addition, US Navy aircraft destroyed more than 270 aircraft in the air and on the ground as well as strafed ground targets and port facilities within the atoll, causing untold damage. The US Navy lost twenty-five aircraft and aircrew either taken prisoner or killed in action, and it lost eleven men on board *Intrepid*, with that carrier out of service for seven months.

The US Navy returned to attack Truk Lagoon again on April 30 and May 1, 1944. After the April carrier attacks, the atoll's garrison was bypassed. Strategic planners in the United States chose to strangle the Truk garrison rather than sacrifice Allied lives in a costly invasion of the atoll. Needing an air base in the Caroline Islands, the Navy and Marines invaded Eniwetok, able to do so without the threat of Japanese aircraft from Truk interfering with the operation. In addition, taking Truk

The 499-foot long *Aikoku Maru*, a 10,438-ton armed transport, explodes from a direct hit, which blew off the entire forward half of the ship from the funnel forward. Dublon Island is in the right center of the photo. The ship was carrying troops, and it is estimated that 730 troops and eleven crew lost their lives when aircraft from the carriers *Essex* and *Intrepid* dropped five-hundred-pound bombs on the ship. *US Navy*

SHIPS SUNK IN TRUK LAGOON

Ship Name	Type	Tonnage	Length (feet)	Depth of water (feet)
Aikoku Maru	armed transport	10,438	499	130 to 240
Amagisan Maru	armed transport	7,620	450	140 to 200
Cha-29	No. 28–class submarine chaser	420	161	unknown
Cha-46	No. 1–class submarine chaser	130	85	unknown
Cha-66	No. 1–class submarine chaser	130	85	unknown
Eisen No. 761	tug	300	112	115 to 200
Fujikawa Maru	armed transport	6,938	436	40 to 120
Fujisan Maru	tanker	9,524	492	140 to 200
Funaitzuki	fast transport destroyer	1590	322	115 to 200
Futagami	ocean tug	625	131	40 to 75
Gosei Maru	transport	1,931	272	20 to 120
Hanakawa Maru	armed special transport	4,739	367	50 to 120
Heian Maru	submarine tender	11,614	465	50 to 130
Hino Maru No. 2	gunboat	998	200	10 to 50
Hoki Maru	transport	7,112	449	25 to 160
Hokuyo Maru	transport	4,217	357	150 to 200
Hoyo Maru	yanker	8,691	475	10 to 111
I-169	Submarine B (1)	1785	337	120 to 150
Katsurigisan Maru	transport	2,427	285	200 to 230
Kensho Maru	transport	4862	383	60 to 140
Kikukawa Maru	ammunition supply	3,833	354	100 to 120
Kiyozumi Maru	armed transport	8,614	452	70 to 120
Minsei	converted minelayer	378	134	unknown
Momokawa Maru	transport	3,829	354	80 to 145
Nagano Maru	transport	3,824	344	165 to 210
Nippo Maru	transport	3,764	354	90 to 165
Oite	destroyer (*Kamikaze* class)	1,523	328	180 to 210
Ojima	salvage tug	812	161	150 to 165
Reiyo Maru	transport	5,446	400	170 to 220
Rio de Janeiro Maru	submarine tender	9,626	462	50 to 110
San Francisco Maru	transport	5,831	383	165 to 210
Sapporo Maru	provision/storeship	361	144	60 to 110
Seiko Maru	armed transport	5,385	393	120 to 160
Shinkoku Maru	armed oil tanker	10020	498	36 to 120
Shotan Maru	armed transport	2,829	305	90 to 150
Susuki	Patrol Boat No. 34	935	275	10 to 55
Tachi Maru	transport	1,891	269	100 to 110
Tachikaze	destroyer (*Minekaze* class)	1,345	337	unknown
Thiho Maru	armed transport	2,827	321	175 to 200
Traijun Maru	transport	1,278	unknown	unknown
Tonan Maru No. 3	oil tanker	19,209	534	unknown
Unkai Maru No. 6	armed transport	3,220	331	100 to 130
unknown	4 landing craft	unknown	59	unknown
unknown	gunboat/landing craft	unknown	59	20 to 40
unknown	lighter/water transport	unknown	131	80 to 100
unknown	picket boat/submarine chaser	unknown	124	150 to 175
unknown	transport/inter-Island supply	unknown	98	80 to 100
Yamagiri Maru	armed transport	6,438	439	80 to 100
Yamakisan Maru	special transport	4,776	367	10 to 150
Yubai Maru	transport	3,217	305	75 to 110

Source: "War in Paradise: World War II sites in Truk Lagoon, Chuuk, Federated States of Micronesia."
www.nps.gov/parkhistory/online_books/npswapa/extContent/wapa/paradise/index.htm

was unnecessary because the anchorage at Ulithi Atoll was more desirable. Ulithi is located 850 miles farther west in the Caroline Islands, 420 miles southwest of Guam, and 1,370 miles to Okinawa and the Japanese home islands.

The Japanese military men on Truk were kept in a constant state of readiness for an Allied invasion that never came. American submarines patrolled the sea lanes around the atoll and kept constant watch on its exits. Truk Lagoon had been bypassed, and although there was an occasional submarine supply mission to the atoll, the men there faced a semi-starvation diet.

In subsequent months, marauding Allied ships and aircraft passing Truk Atoll always made the effort to harass the beleaguered garrison. The British mounted a carrier plane attack on the atoll on June 14 and 15 as well. Army Air Force B-24s harassed the atoll's garrison with bombing raids until the end of the war. Relief for the Japanese finally came on September 2, 1945, when Lt. Gen. Shunzaburo Mugikura and Vice Adm. Chūichi Hara, commander of the Japanese Fourth Fleet, arrived aboard the American cruiser *Portland* (CA-33) to surrender the Truk garrison.

SHIPWRECKS OF TRUK LAGOON

Seventy years after the waters settled from ships sinking during Operation Hailstone, Truk Lagoon today is an ideal dive site. The water temperature ranges from 82° to 84°, and the majority of the shipwrecks sit at depths between 120 and 200 feet. Depending upon the time of year and currents in a particular dive location, visibility ranges from twenty-four to one hundred feet. This means divers can enter the water and soon see their destination beneath them.

This aerial photo gives an idea of how target rich the Truk Lagoon anchorages were when the planes from Task Force 58 attacked on February 16 and 17, 1944. *US Navy*

The peaceful waters of Truk (Chuuk) Lagoon belie the destructive February 1944 aerial attacks on Japanese shipping that occurred here. *NPS Submerged Resources Center*

Right: A Japanese aircraft cockpit, its instrument panel stripped of gauges, sits in the hold of *Fujikawa Maru*. This 437-foot-long wreck is only 40 feet below the surface at its top, bottoming out at 115 feet, and is easily accessible to all levels of divers. It is considered one of the best dive spots in Truk Lagoon. *NPS Submerged Resources Center*

Typically, most dive operators will visit fifteen or so wrecks, with five of them more than 400 feet long (*Heian Maru*, 520 feet long; *Shinkoku Maru*, 500 feet; *Rio de Janiero Maru*, 461; *Fujikawa Maru*, 439; and *Yamagiri Maru*, 437). In addition to the merchant ships, warships include the destroyer *Fumizuki* and the submarine *I-169*. There are also a number of aircraft wrecks, including a Mitsubishi G4M "Betty" twin-engine bomber; a large, four-engine Kawanishi H8K "Emily" flying boat; a Yokosuka D4Y "Judy" dive-bomber; a number of Mitsubishi A6M Zeros in relatively shallow water; and a number of aircraft fuselages in the holds of some of the shipwrecks.

"When I was a kid, most of my friends were reading comic books while I was reading books about the famous German ships that were sunk during World War II. I just had an interest from day one," said Dan E. Bailey, diver and author of *World War II Wrecks of the Truk Lagoon*. "When I took a job offer in Kwajalein in the Marshall Islands, which was also a Japanese naval base, the day I arrived I got lined up with the people who liked to fish. I was down at the docks a lot and was seeing guys heading out to scuba-dive the shipwrecks. I started looking at some of the artifacts they were bringing back and then their underwater photos, and that just fit in with my interests. So I learned how to dive. Some five thousand dives and sixty trips to Truk later, that's where I am."

According to Bailey, *San Francisco Maru* is the most popular shipwreck dive in Truk Lagoon. Each of the dive guides have the locations of the wrecks figured out based on lining up edges of islands or a tree on the shore, and they throw a small

anchor overboard while the live-aboard dive boats have a sub-surface buoy that they tie onto. Dives to each of the wrecks are made with a guide.

San Francisco Maru is a 385-foot-long passenger/cargo ship, built in 1919, that sits upright on the bottom of the lagoon. The ship's configuration is mast-kingpost-funnel-mast, with four holds and a three-inch deck gun mounted at the bow. Depth to the deck is 165 feet, and the bottom lies at 210 feet. Most single-tank (eighty-cubic-foot) air dives give approximately fifteen minutes of dive time at the main deck level. "On this type of dive, you're generally swimming across the deck, looking down into the hold, going through the bridge super structure, working your way down the length of the wreck as far as you can, and just looking down to see what's there," said Bailey. "Going down into the holds, you're obviously going deeper, which cuts down your time. A lot of the diving these days is special-ized—mixed gas, rebreathers, et cetera—and you are able to spend almost an hour with the right equipment."

The exciting part of diving on *San Francisco Maru* is the amount of war materiel that is stored in the ship's holds. "Mines, different kinds of bombs, different kinds of armament, trucks, battle tanks, lanterns. It's filled with munitions of every type imaginable, including torpedoes and depth charges," Bailey said.

Although deterioration of *Fujikawa Maru* is progressing rapidly, it, too, is a popular wreck to visit. The ship is 439 feet long and configured mast-kingpin-funnel-kingpin-mast, with six-inch cannon on both the bow and stern. It has a fantastic bridge to swim through. The bow and stern cannon are coral encrusted, making them very photogenic, and a large amount of ammunition is stacked around the wreck.

Fujikawa Maru's deck is only forty feet below the surface, with the bottom of the holds at ninety feet. Depth to the bottom is 120 feet. The No. 1 hold contains aircraft engines, ammunition of various calibers, and machine guns. Hold No. 2 is home to four disassembled aircraft and wings, while hold No. 3 is empty. Hold No. 4 is empty as well; however, this was the site where an American aerial torpedo blew a hole in the ship's bottom. The remaining two holds were salvaged by the Japanese during the war years, although a large number of bottles, china, and other equipment can still be seen.

One of the rarest wrecks, and a ship that was not sunk during Operation Hailstone, is the fleet submarine *I-169*. The 369-foot-long craft was built at the Mitsubishi shipyard in Kobe and delivered on September 28, 1935, as *I-69*, later re-designated *I-169*. This submarine was patrolling Hawaiian waters on December 7, 1941, its mission to retrieve the midget submariners and return them to Japan. At anchor at Truk Lagoon on April 4, 1944, it dived when B-24 Liberators were approaching to bomb the atoll. As the submarine descended, a valve stuck open, flooding the control room. Unable to surface, a rescue attempt was mounted but ultimately failed. The Japanese then depth charged the wreck to destroy it in an attempt to prevent the Allies from salvaging the sub in the future.

In 1971, the wreck was located at a depth between 130 and 150 feet of water. Divers entered the sub, and film footage from the interior showed many of the crew's remains. Money was raised in Japan to recover the skel-etons and give them a proper Shinto funeral. The submarine's complement was seventy, yet seventy skulls and bones for another thirty personnel were recovered—most likely repair workers who were in the wrong place at the wrong time. Said Bailey:

> To be able to dive a Japanese World War II submarine is just spectacular. I don't know that there are more than two or three in the whole world that you can dive on, and I think this is much shallower than any of the others. The wreck

The cockpit of another Mitsubishi A6M Zero fighter in the hold of *Fujikawa Maru*. The ship was carrying a large load of aircraft components, from fuselages and wings to propeller blades, cowling, and fuel cells. The A6M's instrument panel is unique to that series of fighters, and about half of the instruments remain in this cockpit. *NPS Submerged Resources Center*

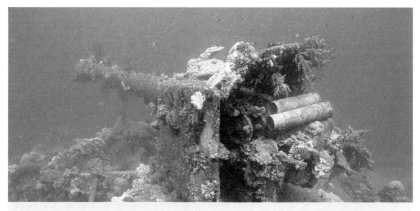

Six-inch deck gun on *Fujikawa Maru*, with spent shell casings stacked on top of the breech. *NPS Submerged Resources Center*

has deteriorated a lot—the depth charging did substantial damage—but you see the torpedo tubes and the torpedoes are actually showing. The conning tower and all of that structure has fallen to the side, and you get a chance to look at all of it and see the entrances into the torpedo room and the bridge. It's pretty interesting.

My favorite warship to dive in Truk is the destroyer *Fumizuki*. It was found later than most of the wrecks by a Japanese man who had a better location than anybody else. It is a virtually intact destroyer.

Fumizuki is a 320-foot-long *Mutsuki*-class destroyer with a thirty-foot beam and a draft of 10 feet. The ship was powered by four boilers driving twin-geared turbines producing 38,500 shaft horsepower, which could move the destroyer through the water at 33.5 knots. Its main armament was four 4.7-inch/45-caliber guns, six 24-inch torpedo tubes (ship's capacity: ten torpedoes), and eighteen depth charges. It was completed on July 3, 1926, and converted to a fast transport in 1941. Most of the armament was retained during the conversion; however, one of the triple torpedo tube launchers was removed.

"The wreck is at a reasonable depth and all of the big guns, the torpedo tubes, the depth-charge throwers, and roll-off racks are still on board. It is just fabulous," said Bailey. "It's about a hundred and forty feet to the bottom, so you can do a nice half-hour dive on that one with just normal air. When you dive that deep, you generally do fifteen to twenty minutes of decompression just to be safe."

Three bombs sent *Nippō Maru* to the bottom of Truk Lagoon. This cargo ship was converted to carry fresh water in hold No. 2. The wreck is 354 feet long, has a 50-foot beam, and is configured mast-kingpost-funnel-mast. It lies in 165 feet of water with a 25° list to port. The ship was fitted with bow and stern guns, but they were removed before the February 1944 attack on Truk Lagoon.

Hold No. 1 carries mines, barrels, and a variety of ammunition, and on the main deck, alongside hold No. 2 on the port side, is a Japanese tank. On the starboard side, next to hold No. 4, is a group of small howitzers, and a couple of large-caliber gun tubes sit near hold No. 5. There are more artillery pieces in No. 5 as well.

A Japanese tank carried as deck cargo on the 354-foot-long *Nippō Maru*. The wreck of this ship is close to the eastern shore of Dublon Island and sits upright, although listing to port. She rests on the bottom between 140 and 170 feet below the surface. *NPS Submerged Resources Center*

There are more than two dozen shipwrecks and another half-dozen aircraft wrecks to see within the atoll. From Jacques Cousteau's 1971 documentary *Lagoon of Lost Ships* to divers who entered the waters of Truk Atoll yesterday, all agree that this lagoon has the best wreck diving anywhere in the world. The variety of ship sizes and types, the variety of cargoes, and the water temperature and clarity all combine to draw divers and historians to this premier graveyard of World War II–era vessels.

BIKINI ATOLL BATTLEFLEET

BETWEEN 1944 AND TODAY, ESTIMATES ARE THAT the world's superpowers built more than 128,000 nuclear weapons—hydrogen bombs, fission bombs, fusion bombs, clean bombs, dirty bombs, neutron bombs, strategic nuclear weapons, tactical nuclear weapons, intercontinental ballistic missiles with MIRVs (multiple independent reentry vehicles), air- and sea-launched nuclear-tipped cruise missiles, artillery shells that delivered nuclear destruction, and little nuclear weapons that could be toss-bombed from aircraft as small as an A-4 Skyhawk. The Americans and the Soviet Union developed 98 percent of all these weapons—nuclear bombs for any and all engagements with the enemy, no matter how big or how small.

Weapons tests were conducted in the atmosphere, belowground, and underwater. The US military even cooked off a nuclear bomb only five hundred miles off the coast of San Diego as part of Operation Wigwam, and gamblers in Las Vegas could see mushroom clouds on the horizon as bombs exploded at the Nevada Test Site.

The United States conducted 235 atmospheric nuclear weapons shots from 1945 to 1962 (the year of the last US atmospheric test), when aboveground, underwater, and space-based tests were halted with the August 5, 1963, signing of the Limited Nuclear Test Ban Treaty.

EXPANDING AMERICA'S NUCLEAR KNOWLEDGE

At the end of World War II, only three nuclear devices had been exploded, and there was a thirst for knowledge about what exactly these bombs could do and how they could be employed as weapons. The first atomic bomb was exploded on the desert, southwest of Socorro, New Mexico. That bomb test, codenamed "Trinity," ushered in the nuclear age. The second device, codenamed "Little Boy," was exploded over Hiroshima, Japan, on August 6, 1945, with devastating effect. A follow-up bombing three days later saw "Fat Man" explode over the Japanese city of Nagasaki.* More than 150,000 Japanese died instantly when the bombs exploded, and the lasting effects from the radiation have impacted the cities' population in the decades since.

Unable to capture much information about the bombings of Hiroshima and Nagasaki, the US military needed to see how an atomic bomb behaved and observe it in a controlled environment. It was decided that a test would be conducted, and it would be given the code name Operation Crossroads. The bombs for the Crossroads tests were similar in construction to the Fat Man bomb.

The first objective of Operation Crossroads was to study the effects of an atomic bomb upon ships, military equipment, and other materiel. The second objective was to study the behavior of an atomic explosion, ranging from the initial blast to its heat and radiation characteristics to wave action generated from an underwater explosion. Among many other separate investigations, the electromagnetic pulse from the explosions was monitored at points around the Pacific Rim—in Alaska, Hawaii, Manila, on the US West Coast, and as far away as Germany. Radars and radios were placed in various locations, including aboard ships of the target fleet, to see how they would be affected when the bomb went off. Lessons learned from the tests would be incorporated into future warship and equipment designs.

Three "shots" (Able, Baker, and Charlie) were planned at a now-infamous remote location in the Marshall Islands, known as Bikini Atoll. The atoll is located between Guam, which is 1,390 miles to the west, and the Hawaiian Islands, which are 2,511 miles to the east. It comprises twenty-three islands that cover 3.4 square miles of landmass and surround a 229-square-mile lagoon.

The Crossroads tests were conducted under the auspices of Joint Task Force 1 (JTF-1). The task force, formed on January 11, 1946, brought together Army, Navy, and civilian contractors and scientists, managed the diverse groups of personnel and equipment required for the tests, and acted as a clearing house for the distribution of all reports generated about the tests and their aftermath. On site, the staff of JTF-1 conducted operations from the amphibious forces command ship *Mount McKinley* (AGC-7). To support the operation, thirty-seven thousand US Navy men and six thousand Army Air Force, scientific, academic, and contractor personnel were gathered. A flotilla of 150 ships supplied bunks, workshops, and laboratories for the men. In addition, JTF-1 used locations on nearby Enewetak and Kwajalein Atolls as support bases. For the tests, the men of JTF-1 were pulled back ten nautical miles east of Bikini Atoll—thought to be a safe distance from the blast.

It is worth noting that the small, two-piece swimsuit—using only thirty square inches of fabric and developed by French designer Louis Réard—was publically announced on July 5, 1946, and was named for the Bikini Atoll tests.

While the target ships streamed into the atoll, native Bikinians were relocated from their homes to nearby Rongerik Atoll. Bikini Atoll was completely evacuated by March 7, 1946, when the last of its 167 inhabitants were delivered to Rongerik by *LST-1108*. The Bikinians were unhappy with living conditions on Rongerik Atoll and, after protracted negotiations, were moved on November 3, 1948, approximately

---

* Both the Trinity test article and Fat Man were implosion-type bombs with plutonium cores, while Little Boy was a gun-type weapon. In the latter type, a conventional explosive inside the bomb ignites to fire a hollow uranium pellet down a barrel onto a cone-shaped cylinder of uranium-235, and the resulting impact creates fission.

four hundred nautical miles south to Kili Island, which is only a short distance to Jaluit Atoll.

For Operation Crossroads, the US Navy began assembling more than ninety vessels that were to serve as test subjects for both of the twenty-three-kiloton shots (equivalent to twenty-three thousand metric tons of TNT). Battleships *Nevada* (BB-36), *New York* (BB-34), and *Pennsylvania* (BB-38) were joined by the aircraft carriers *Saratoga* (CV-3) and *Independence* (CVL-22), the Japanese battleship *Nagato* and cruiser *Sakawa*, the German cruiser *Prinz Eugen*, and a variety of cruisers, destroyers, submarines, landing craft, and many others. The ships were moored in the target area on the eastern side of the atoll in the water between Bikini and Eneu Islands, directly west of Iomelan Island.

ABLE TEST

Ninety-three target vessels were gathered for the first shot of Crossroads, set for July 1, 1946. To record the blast, the Navy assembled an air fleet of forty-three aircraft, including six Grumman F6F drones, sixteen F6F drone controllers, six F6Fs used to photograph the explosion, two PBM Mariners for radiological reconnaissance, three more for air-sea rescue, and four for photoradiometry. The fleet also included eight TBM Avenger torpedo-bombers, which would provide additional photography and air-sea-rescue assets, and four Avengers, which controlled drone boats that sailed through the test area taking water and air samples. Six PB4Y-2 Privateers also flew weather-reconnaissance and search-and-rescue missions for the operation.

In support of the Able shot, the Army Air Forces gathered seven B-29s (one for aerial command, one to drop the bomb, two to drop air pressure gauges, and three spares); eight F-13s (photo-reconnaissance versions of the B-29); two C-54s for additional photography; two F-13s for radiological reconnaissance; ten drone B-17s to take samples as the mushroom cloud formed and during the explosion's aftermath; six drone-controller B-17s; three weather-reconnaissance WB-29s; a mix of thirty C-46 and C-54 transports; a C-54 and a B-29 for observation, radio broadcast, and press; and, finally, two SB-17 air-sea-rescue planes.

On June 30, 1946, the majority of the support ships evacuated the harbor in anticipation of the next day's test. Around midnight, the bomb was prepared at Kwajalein and loaded into B-29 44-27534, *Dave's Dream*. Early in the morning of July 1, at 5:40 a.m. local time, *Dave's Dream* was ordered into the air by the commander of JTF-1, followed by the Navy's fleet of observation planes. By 8:00 a.m., seventy-nine aircraft were in position in the skies around Bikini Atoll. *Dave's Dream* made one test pass over the target ships, and at 8:50 a.m. it was ordered to commence the bomb run.

The Able test bomb detonated 520 feet above the surface of the atoll, but between 1,500 and 2,000 feet off target. The explosion occurred near the Japanese cruiser *Sakawa* and the submarine *Skate* (SS-305), approximately 6,300 yards southwest of Bikini Island. An unnamed Army doctor observing the explosion from a Navy PBM Mariner patrol plane twenty miles from the blast described the explosion in *Operation Crossroads 1946* (Final Report, Defense Nuclear Agency, 1984):

> At 20 miles it gave us no sound or flash or shock wave. . . . Then, suddenly we saw
> it—a huge column of clouds, dense, white, boiling up through the strato-cumulus,

looking much like any other thunderhead but climbing as no storm cloud ever could. The evil mushrooming head soon began to blossom out. It climbed rapidly to 30,000 or 40,000 feet, growing a tawny-pink from oxides of nitrogen, and seemed to be reaching out in an expanding umbrella overhead. For minutes the

Two different angles of the Baker Test explosion at Bikini Atoll. The pagoda superstructure of the battleship *Nagato* makes the ship easy to identify; to the right of her is the battleship *Nevada*. The after-cloud of the explosion is beginning to fan out, and the five-hundred-foot-tall base surge from the explosion is rushing toward the ships surrounding the zero point (the exact location of the bomb's explosion). *Author's collection and NPS Submerged Resources Center*

cloud stood solid and impressive, like some gigantic monument over Bikini. Then finally the shearing of the winds at different altitudes began to tear it up into a weird zigzag pattern.

Radioactivity was negligible, and the majority of the ships were deemed safe to board the following day. What the returning sailors found was that ships within five hundred yards of the air burst had sunk immediately or been seriously damaged. The blast immediately sank destroyers *Anderson* (DD-411) and *Lamson* (DD-367), attack transports *Carlisle* (APA-69) and *Gilliam* (APA-57), and the Japanese cruiser *Sakawa*. Another six ships were seriously damaged, eight were considered heavily damaged, nine were moderately damaged, and forty-three suffered negligible damage. Fourteen aircraft on carrier decks were completely destroyed, and thirty planes were seriously damaged.

BAKER AND CHARLIE TESTS

For the Baker test on July 25, 1946, sixty-eight target vessels were positioned around Landing Ship Medium 60 (*LSM-60*) in the blast area and another twenty-four landing and other small craft were beached on Bikini Island. The bomb for the Baker shot was prepared at Kwajalein and loaded onto *LSM-60* at Bikini Atoll. It was then lowered ninety feet under the surface of the atoll and suspended in a watertight housing made from the conning tower of the scrapped submarine *Salmon* (SS-182). The bomb was detonated at 8:35 a.m., yielding the anticipated twenty-three kilotons.

The same Army doctor quoted for the Able shot described this detonation:

> The flash seemed to spring from all parts of the target fleet at once. A gigantic clash—then it was gone. And where it had been now stood a white chimney of water reaching up and up. Then a huge hemispheric mushroom of vapor appeared like a parachute suddenly opening. . . . By this time the great geyser had climbed to several thousand feet. It stood there as if solidifying for many seconds, its head enshrouded in a tumult of steam. Then slowly the pillar began to fall and break up. At its base a tidal wave of spray and steam rose to smother the fleet and move on towards the islands. All this took only a few seconds, but the phenomenon was so astounding as to seem to last much longer.

The target ships were bathed in a sea of radioactive saltwater, and a wave from the blast washed over the small, low-lying islands between Bikini and Eneu Islands. *LSM-60*, the ship at surface zero—the point of the explosion—was instantly vaporized, and seven other ships soon sank. These included the battleship *Arkansas* (BB-33), aircraft carrier *Saratoga* (CV-3), submarines *Apogon* (SS-308) and *Pilotfish* (SS-386), oiler *YO-160*, floating dry dock *ARDC-13*, and Japanese battleship *Nagato*.

Helicopters were used to retrieve exposed film from Bikini Island, but the area remained radioactively "hot" for another six days. Decontamination of ships began on August 1, and some ships were so hot that sailors could only board them for a matter of minutes before their dosimeters reached capacity. In 1946, 0.1 Roentgen was considered the maximum amount of radiation exposure per day without any long-term health effects.

The aircraft carrier *Saratoga* (CV-3) sank beneath the atoll's surface approximately seven and a half hours after the explosion. Most of the carrier's exhaust stacks were crushed and collapsed onto the flight deck, the majority of aircraft and equipment lashed to the deck was blown overboard, and the hull was severely washboarded from the bomb's pressure. *NPS Submerged Resources Center*

Ten days later, JTF-1 realized that decontaminating ships with water from Bikini Atoll was defeating the purpose, and the decision was made to tow the target fleet to Kwajalein Atoll. Here, ships would be decontaminated with fresh seawater, and off-loading and safe storage of ammunition from the ships could be accomplished as well.

Twelve ships escaped full irradiation and were safe enough to sail back to the continental United States. Eight of the capital ships and two submarines were towed back to the West Coast (Hunters Point Naval Shipyard in San Francisco, California, and Bremerton, Washington) and to Hawaii for further decontamination. Those ships too radioactive or too damaged to salvage were sunk in deep water off Kwajalein Atoll or off the Hawaiian Island of Oahu. Battleship *Pennsylvania* (BB-38) was scuttled off Kwajalein Atoll on February 10, 1948, and *New York* was sunk off Oahu on July 6, 1948, followed by *Nevada* on July 31, 1948.

The third and final blast of Operation Crossroads was to be the Charlie shot. It was planned as a deep-water nuclear detonation, but the test was not conducted. President Harry S. Truman called off the test when engineers could not complete construction of a bathysphere needed to hold the bomb by the deadline for its deployment to the Pacific.

SHIPS SUNK DURING BIKINI NUCLEAR TESTS

Name	Type	Nation	Year Built
Able Test (air drop from B-29), July 1, 1946			
Anderson (DD-411)	destroyer	USA	1939
Carlisle (APA-69)	attack transport	USA	1944
Gilliam (APA-57)	attack transport	USA	1944
Lamson (DD-367)	destroyer	USA	1936
Sakawa	cruiser	Japan	1944
Baker Test (submerged explosion), July 25, 1946			
Apogon (SS-308)	submarine	USA	1943
Arkansas (BB-33)	battleship	USA	1911
Nagato	battleship	Japan	1919
Pilotfish (SS-386)	submarine	USA	1943
Saratoga (CV-3)	aircraft carrier	USA	1925
ARDC-13	auxiliary repair dry dock, concrete	USA	1946
LCT-1114	landing craft, tank	USA	1944
LCT-1175	landing craft, tank	USA	1944
LCVP-10	landing craft, vehicle personnel	USA	1945
LSM-60	landing ship, medium	USA	1944
YO-160	yard oiler	USA	1943
Sunk after Testing			
LCM-4	landing craft, mechanized	USA	1943
LCT-414	landing craft, tank	USA	1942
LCT-1187	landing craft, tank	USA	1944
LCT-1237	landing craft, tank	USA	1944
Prinz Eugen	heavy cruiser	Germany	1938

SURVEYING THE BIKINI FLEET

Decades after the Crossroads tests, a joint US Navy–National Park Service team conducted a survey of the sunken fleet at Bikini Atoll. Retired NPS archaeologist and diver Daniel J. Lenihan, leader of the team, said:

> Part of the reason we were there was to assess not only the archaeological but the recreational potential of these sites so that the Bikinians could use it as a source of income while preserving it.

There were issues that had to do with radiation, and we weren't qualified to be making the scientific determinations about the radiation. We were assessing whether the Bikinians could have people pay to travel there and do guided dives over these ships. There was a list of answers we had to obtain from the work we were doing. Part of it was to determine if they had the workable makings of a National Park site, something that was non-consumptive, that didn't use up the resource while they were doing it. That struck us as a good recommendation to make. That's why our assessment concentrated on ships that would be symbolically important to people, like *Saratoga* and *Nagato*. It is a tremendous diving experience to be on those sites.

We were gauging, as experienced divers, if it was a reasonable thing to do safety-wise, and was it going to be a gratifying diving experience? Of course, the answer to that [second] one is easy—it's definitely a very gratifying experience to any diver.

Lying on the bottom of Bikini Atoll within one mile of the explosion point are eight of the ships sunk during the atomic testing—*Arkansas*, *Nagato*, *Saratoga*, submarines *Apogon* and *Pilotfish*, attack transports *Carlisle* and *Gilliam*, and the concrete-hull yard oiler *YO-160*. The maximum depth in this area of the atoll is approximately two hundred feet, and visibility ranges from sixty to one hundred feet. Underwater visibility would enable a diver to see from one end to the other of most ships except for the battleships and the carrier.

Battleship *Arkansas* was 620 yards from the detonation point, and its hull, along with those of attack transports *Carlisle* and *Gilliam*, was ripped open. *Arkansas* took the full brunt of the explosion, and it is believed it was lifted on end during the blast and slammed down to the bottom of the lagoon when the radioactive water plume dropped back to the surface of the atoll. It had disappeared from the surface of the lagoon by the time the smoke cleared. Of the wreck, Lenihan said, "*Arkansas* looks like a really giant foot came down and jammed it into the silt, inverted." *Arkansas*'s stern was sheared off in the blast, and two of the propeller shafts are missing from the wreck.

The 888-foot-long aircraft carrier *Saratoga* was only 350 yards south-southwest of the center point of the Baker blast at the time of detonation. Like *Arkansas*, it was severely damaged by the force of the underwater explosion. Most of the equipment on deck was blown overboard, and the starboard-side torpedo blister section of the hull—the side facing the blast—is heavily wash-boarded at the center of the ship. The smokestacks are blown over and lying across the deck, and a number of large, deep depressions crisscross the aft section of the flight deck. Said Lenihan:

Saratoga is kind of otherworldly. That ship is nine hundred feet long, and three hundred feet—one-third—longer than the *Arizona*. It's in clear water, and as you descend onto the flight deck it's just an impressive sight to behold. As you start dropping down, your eyes only go to the mast, the superstructure on top of the island. Diving farther down over the command center you start getting more of a sense of the massive size of the ship, since it reaches the limits of your visibility,

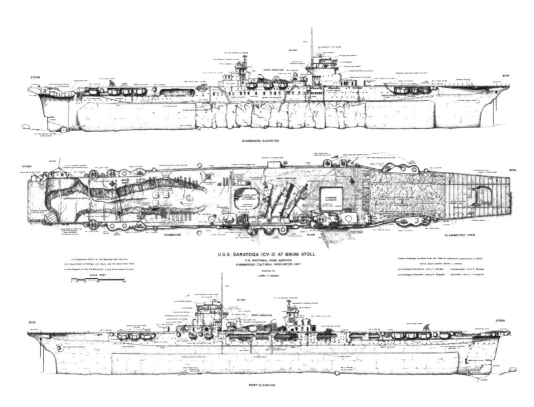

This detailed wreckage map of the *Saratoga*, compiled by Larry V. Nordby from the Submerged Resources Center, shows the damage done to the hull, flight deck, and stacks. Note the items called out on the sea floor next to the carrier. The Army 155-millimeter cannon, mounted to the port flight deck aft of the island, has separated from its mount and survived the sinking, ending up fifty feet farther forward. The dynamics of the blast and subsequent on-rushing water impacted objects of varying densities in different ways—eight-thousand-pound aircraft were blown overboard while a two-hundred-pound coil of loose cable barely moved an inch. *NPS Submerged Resources Center*

fore and aft. The flight deck disappears off in both directions. Rarely do you have that kind of visibility where you can see that far when diving. Hovering over the flight deck is like walking on it when it was afloat. You're able to get a sense of the ship heading in to the gloom a couple of hundred feet in each direction.

The flight deck is at a depth of 95 feet at the bow and 120 feet at the stern. That's the kind of slant the ship lays on. We learned later that many people will swim the length of the flight deck and that will be their entire dive. Down and back is one-third of a mile. It's a fascinating dive. That and just diving around the island are two of the dives that people really enjoy.

If you're swimming down the flight deck, all of the wood that once formed the deck where the planes were taxiing or taking off has gone the way of the Teredo worms [*Teredo navalis*]. What you've got now is little islands of wood around all of the fittings that held the deck in place. The wood preservative soaked into the wood around the deck pins, and that keeps it from being eaten by Teredo worms. You've got thousands of little islands of wood going down the deck.

When you get to the command center, you can swim inside. The room is very interesting with the brass battle shields down over the portholes. They let in just a slit of light. That's a pretty wild feeling to be inside there. Now in the dark looking out to sunlit water— it's like the ship is underway.

Descending into the hangar deck, divers come down over a set of bombs that have been laid out across the floor. There's a lot of live ammunition, live shells and bombs, all throughout the ship. One of the questions we had to address was the large amount of live explosives; what kind of problem does it present? Essentially, our message was that, in just about all cases, this stuff was not a safety issue to anyone in their right mind. You could find a way of making it a problem, perhaps, but the ordnance is not going to jump up and do anything to passing divers.

Inside the hangar deck there were three or four Curtiss Helldiver dive-bombers, which, when we were there at least, you could see the instrument panels and the gauges were intact. There was light coming down through the huge elevator shaft that came down from the flight deck. It's just a very eerie feeling. In a sense,

Talk about a product endorsement. Archaeologist and diver Daniel Lenihan (chief of the Submerged Resources Center at the time this photo was taken) examines an intact light bulb in a passageway ceiling fixture below decks on *Saratoga*. The bulb, brand unknown, survived a nuclear explosion and the pressure change of at least six atmospheres when it sank to the bottom of the lagoon. *NPS Submerged Resources Center*

you know you're underwater, but you're looking at what seems to be an operational flight deck. These planes looked as if they could be taken up to the flight deck and take off from there. You had a feeling you're swimming back into the past.

The Japanese battleship *Nagato* is a massive ship. It was the lead ship of *Nagato* class, which comprised *Nagato* and *Mutsu*. These battleships each displaced 32,720 tons and were 708 feet long, with a beam of 95 feet 3 inches and a draft of 29 feet 9 inches. *Nagato*-class ships were capable of making twenty-six knots and were each crewed by 1,333 officers and sailors. Each vessel's main armament consisted of eight sixteen-inch guns mounted in pairs in four turrets with twenty

The cockpit of an SBF-4E Helldiver (BuNo 31850) on the hangar deck of *Saratoga* shows that most instruments in the panel are intact, yet the pilot's seat has come loose from its mountings. Aviation enthusiasts would love to see the aircraft on *Saratoga's* hangar deck recovered. Today there are less than ten surviving Helldivers, and only one is currently flying. *NPS Submerged Resources Center*

Larry Murphy, former National Park Service archaeologist and chief of the Submerged Resources Center, swims past one of *Nagato's* four propellers. The ship rests, inverted, at a depth of 170 feet. All four of the ship's sixteen-inch gun turrets are connected to the hull and did not separate when the battleship rolled inverted. *NPS Submerged Resources Center*

140-millimeter cannon as secondary armament and four 76-millimeter antiaircraft guns. There were eight twenty-one-inch torpedo tubes as well.

During the Able test, *Nagato* was only lightly damaged, as its distance from the explosion point was farther than expected due to the bomb being dropped off target. Following the Baker shot, *Nagato* remained on the surface for a few days until water seeping into the hull exceeded her capacity to float. At the first opportunity to survey the target fleet, *Reclaimer* (ARS-42) sailed past *Nagato* on July 27 at 4:30 p.m. and saw that the battleship was settling by the stern. The next morning, the latter had an 8° list to starboard, and by 5:00 p.m. on July 29 its main deck was awash. On July 30 at 7:00 a.m., *Nagato* capsized and sank to the bottom of the lagoon, landing inverted.

Lenihan described the wreck:

The *Nagato*. That's really a wild one. Like the *Saratoga*, it's visually stimulating. You're about one hundred thirty feet down when you reach those intact screws and you look off into the distance over the massive hull as it melts away underneath you. Divers get one great view of the screws, and then they have to go into a deep dive to really see the rest of the ship. You have to go over and down the

Jerry Livingston, former National Park Service illustrator responsible for many of the hand-drawn maps and shipwreck diagrams generated by the Submerged Resources Center, floats above the submarine *Pilotfish* (SS-386) while studying the wreck. On the cigarette deck, behind the periscope shears, the twin twenty-millimeter antiaircraft cannon point toward the surface. *Pilotfish* was moored 363 yards from the center point of the bomb's explosion. *NPS Submerged Resources Center*

side of the ship to get to the main deck, and then swim under it to get to the guns. It is semi-dark there and as your eyes adjust, stuff just starts popping out at you.

When we dove the wreck, we were using air. It's 180 feet deep, which is actually a pretty benign depth at Bikini because it's warm, light, and clear. That's significant as that makes the depth less foreboding and less challenging. You can really enjoy the site.

When you're diving at that depth, there's just no way you're not going to have your imagination stirred a bit. Diving underneath the *Nagato* and seeing those massive gun barrels, immediately reminded me of the muzzles of the guns on the *Arizona*. Those guns are the payload of a battleship—everything else is designed to get the guns into firing position. Visitors to Pearl Harbor never get to see the *Arizona*'s guns except on film. They have a very dramatic effect. There is a similar effect on the *Nagato*.

Essentially, when you're looking down the muzzles of the guns on *Nagato*, the connection for me with the *Arizona* was a pretty direct one. The *Nagato*'s bridge is where Admiral Yamamoto was when he heard the Pearl Harbor attack code: *Tora, Tora, Tora*. You can turn the other way and see the muzzles of the guns for which the shells were made that were converted into bombs that were dropped at Pearl Harbor and sank the *Arizona*. It's a pretty direct connection for me.

Two submarines lie at the bottom of the atoll as well. Both *Apogon* (SS-308) and *Pilotfish* (SS-386) sit within the one-mile radius of the Baker blast, with *Apogon* closest to *Saratoga* and *Pilotfish* near *Arkansas*. *Apogon* was 850 yards from the epicenter of the blast and submerged one hundred feet below the surface when the bomb was exploded. Divers attempted to salvage the sub in 1946 but were unsuccessful. Today she sits upright on the bottom in 160 feet of water and is a very popular wreck to dive.

Pilotfish is deeper, at 175 feet, and her hull torn open; pressure from the blast wrapped her external skin around its frame. The upper deck has collapsed, and she lies on the bottom with a 30° to 40° list to starboard.

Having survived the Able and Baker tests at Bikini Atoll, the German cruiser *Prinz Eugen* was towed to Kwajalein Atoll. During a storm, the cruiser capsized and sank in the lagoon at Enubuj Island. The depth of the wreck varies from 25 to 110 feet. Two of the ship's three screws and her rudder are visible in this view; a small fishing trawler, seen in the background at left, has sunk in near the cruiser. *John Voss*

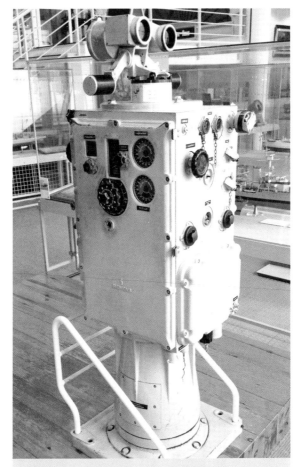

Before *Prinz Eugen* was sent to Bikini Atoll, a number of her systems were removed and studied, including this torpedo director from the cruiser, which is displayed at the German Maritime Museum in Bremerhaven, Germany, on the North Sea. *Prinz Eugen* was built at Kiel, approximately sixty miles to the east on the Baltic. *Nicholas A. Veronico*

The German cruiser *Prinz Eugen* played a major role in the Battle of the Denmark Strait, scoring three hits on the British battlecruiser HMS *Hood*. At the end of the war, it was surrendered to the British and subsequently turned over to the US Navy. The Navy, in turn, commissioned the cruiser as miscellaneous vessel IX-300, keeping the name *Prinz Eugen*. Heavily damaged in the Baker shot, after repeated attempts at Bikini Atoll to clean her, the cruiser was towed to Kwajalein Atoll on August 20, 1946.

The cruiser was too radioactive for work crews to spend much time on board. Its leaks exceeded the capacity of the pumps, and the ship was beached on Carlson Island (today Ejubuj Island). On December 22, 1946, she rolled over and sank in water ranging from 25 to 110 feet deep. Today, the cruiser's center and starboard

One of *Prinz Eugen*'s screws was recovered from the wreck, along with twelve feet of the prop shaft, and transported to the German Naval Memorial at Laboe, Germany, on the Baltic Sea. The bronze propeller forms the centerpiece of a garden that contains a model of the *Prinz Eugen* and other artifacts. *Nicholas A. Veronico*

propellers and rudder are visible above the surface, and divers can swim underneath the upturned hull to see the gun turrets, antiaircraft armament, and broken parts of the ship's superstructure lying on the sand beside the ship. *Prinz Eugen*'s B turret unseated when she turned turtle, while A turret—its guns removed before Operation Crossroads—remains in place.

Parts of *Prinz Eugen* have been returned to Germany. At the German Maritime Museum in Bremerhaven, on the North Sea, is a torpedo director from the historic cruiser. One of the cruiser's massive propellers and twenty feet of the shaft were recovered from the wreck at Kwajalein Atoll and were transported to the German Naval Memorial at Laboe, on the Baltic Sea. The propeller and a model of the cruiser are the centerpiece of the courtyard leading to the memorial. The ship's bell is also on display at the Washington Navy Yard, Washington, DC.

Sixteen ships sank in Bikini Atoll as a direct result of the Able and Baker tests, while more than a dozen additional ships and landing craft were sunk to minimize the damage that could be done by the lingering effects of radiation. The ships of Operation Crossroads are an underwater museum of World War II naval power that presents ships from both sides of the conflict.

"THEN THEY DID IT! ON AUGUST 15, 2006, MY BROTHER JOHN AND I RECEIVED AN EMAIL FROM THE SHIP THAT THEY WERE NINETY-FIVE PERCENT CERTAIN THEY HAD FOUND THE *GRUNION*. IT WAS JUST UNBELIEVABLE! . . . WHEN THE SUB WENT DOWN, IT STRUCK A RIDGE ON AN UNDER-WATER VOLCANO, THEN SLID THREE-QUARTERS OF A MILE TO ITS FINAL RESTING PLACE. THAT TRACK IN THE SEA FLOOR WAS SO OUT OF PLACE; IT WAS ESSENTIALLY THE DEBRIS TRAIL WE WERE LOOKING FOR."

—BRUCE ABELE, SON OF USS *GRUNION* SKIPPER LT. CMDR. "JIM" ABELE

№ 4

NO LONGER
HIDDEN

USS *GRUNION* (SS-216)—LOST WITH ALL HANDS

"FOR SOME REASON WE ALWAYS CALLED HIM 'JIM,'" Bruce Abele said in his soft-spoken New England accent. One could hear the caring and fondness filling his voice as he reflected on his father, Lt. Cmdr. Mannert L. "Jim" Abele, skipper of the World War II *Gato*-class submarine USS *Grunion* (SS-216). Bruce and his two brothers, John and Brad, were only kids when their father and his sixty-nine shipmates went missing during the dark, early days of World War II.

Bruce is now in his eighties, and, like his brothers, he is a man of many accomplishments. In Bruce's voice one can hear the sense of self-imposed responsibility—responsibility to the families of each man aboard the

One of the few surviving photos of Lt. Cdr. Mannert L. "Jim" Abele in uniform. Before taking command of *Grunion*, Abele commanded the submarine *R-13*. He was a very experienced submarine captain and was highly regarded by his men and his peers. The World War II *Sumner*–class destroyer *Mannert L. Abele* (DD-733) was named in his honor. *Courtesy of the Abele family*

sub, a mission that continues to this day, more than seventy years after *Grunion* went missing. That determination led to one of the Abele brothers' greatest personal achievements: finding the final resting place of their father and his crew, thereby bringing closure not only for themselves, but for the families of *Grunion*'s crew.

Yet Bruce and his brothers did not find the sub alone. Solving the mystery of *Grunion* was a journey that took time, technology, resources, and, above all else, the generosity of others and a cadre of collaborators from around the globe.

GRUNION WAR PATROL: ENGAGING THE ENEMY AT KISKA

Margaret N. Hooper, wife of Adm. Stanford C. Hooper, broke the traditional bottle of champagne across the bow of *Gato*-class submarine USS *Grunion* (SS-216) on December 22, 1941. The submarine slid down the ways at the Electric Boat's Groton, Connecticut, shipyard as invited guests looked on. Lieutenant Commander Abele had previously commanded the submarine *R-13*. His new boat, *Grunion*, was commissioned at Submarine Base New London, Connecticut, on April 11, 1942, and for the next six weeks the skipper put his newly built vessel through its paces. A test dive to 315 feet and torpedo trials in Long Island Sound were included.

The Abele brothers—from left to right, John, Bruce, and Brad—always wondered what happened to their father, his shipmates, and their submarine. Each had successful careers and raised families, and it was only in later life that time, technology, resources, and opportunity to collaborate with others all came together enabling them to solve the mystery of *Grunion*'s final resting place. *Courtesy of the Abele family*

On May 24, *Grunion* sailed for the Pacific theater of war. As it zigzagged along the eastern seaboard, Abele worked to identify each vessel he came across. On May 29, *Grunion* passed from the Atlantic Ocean into the Caribbean Sea, cruising submerged past Mayaguana Island and through the Windward Passage.

En route to the Panama Canal on May 31, *Grunion* came across sixteen survivors from the Army Transport USAT *Jack*. The ship had been torpedoed by the German *U-558*, and the survivors floated in their damaged lifeboat for 108 hours

The *Gato*-class submarine *Grunion* (SS-216) slid down the ways at Electric Boat's Groton, Connecticut, yard on December 22, 1941, just weeks after the Japanese attack on Pearl Harbor. She was lost in the Aleutian Islands eight months later. *US Navy*

until they were rescued by *Grunion*. One other raft was known to have cast off from the sinking transport, but after a dedicated search, Abele was forced to continue his journey. Those men were never heard from again. Steaming across the Caribbean Sea, *Jack*'s survivors were landed at the submarine base at Colo Solo, Panama Canal Zone, on June 3. Among those rescued were the *Jack*'s chief mate; Peter Korb; first engineer, George F. Drew; and Lt. j.g. A. O. Lund, chief of the ship's Naval Armed Guard.

The submarine transited the Panama Canal on June 6, 1942, and crossed the eastern Pacific Ocean without incident. A couple of ships were spotted as were a

First Day Cover commemorating the commissioning of the submarine *Grunion*, April 11, 1942. *Author's collection*

Grunion is seen during trials on Connecticut's Thames River near the Navy's New London submarine base. The submarine was 311 feet 9 inches long, with a beam of 27 feet 3 inches. She could travel more than eleven thousand nautical miles at ten knots while surfaced. *US Navy*

few US patrol planes. The voyage took seventeen days with the boat arriving at Pearl Harbor on the morning of June 20. *Grunion* sailed into Pearl Harbor, past the overturned battleship *Oklahoma* and the wreck of the once-mighty *Arizona*, sitting on the bottom of the harbor. The crew could see salvage workers attending to each ship. *Grunion* was tied up at the undamaged submarine base, the reason for her mission visible across the channel.

At Pearl Harbor, *Grunion* underwent a rigorous training schedule to get the boat and her crew into top fighting condition. The submarine would conduct torpedo attack training using Mk 14 exercise torpedoes modified so they could later be retrieved and used again. Once torpedo training was done for the day, the tables were turned and the submarine served as a target for US destroyers attempting to track the sub using sonar and as a target for depth-charge runs. *Grunion* used this part of the cat-and-mouse game to perfect its ability to hide from its pursuers.

The weekend of June 26–28 was probably spent on one last liberty call as *Grunion* got underway at 9:05 a.m. local time on June 30. *Grunion*'s ultimate destination would be the Aleutian Islands, specifically in the area around Kiska Island. The Japanese had occupied both Attu and Kiska on June 3 and 4, simultaneous to their attack on Midway Atoll.

With Oahu passing by the stern that afternoon, *Grunion* attacked a training target in Kauai Channel, firing ten rounds of 3"/50 from the deck gun, and some .50-caliber machine gun fire raked the target as well.

Official Navy understanding of *Grunion*'s activities for the period July 15–30 is very sketchy. Radio transmission was poor, and there were only four brief and somewhat confusing communications from the sub, some of them being repeats.

It is known that on July 15, *Grunion* attacked and sunk what was believed then to be three destroyers. (After the war it was determined that they were sub chasers and that only two were sunk.) *Grunion* made several other attacks, but the results were unconfirmed.

In his memoir *Submarine*, Capt. Edward L. Beach, communications officer on *Trigger* (SS-237), reported decoding the following *Grunion* message:

ATTACKED TWO DESTROYERS OFF KISKA HARBOR NIGHT PERISCOPE SUBMERGED X RESULTS INDEFINITE BELIEVE ONE SANK AND ONE DAMAGED X MINOR DAMAGE FROM COUNTERATTACK TWO HOURS LATER ALL TORPEDOES EXPENDED AFT.

It was not dated, so its relevance is difficult to assess. Beach ends his description of *Grunion*'s message with the comment that it decoded perfectly up to a point and then became a jumble.

That afternoon, July 30, *Grunion* was ordered to the American naval base at Dutch Harbor on Unalaska Island. In her acknowledgment of the order, *Grunion* gave the following report:

WITH VISIBILITY 400 YARDS HEARD ECHO RANGING SHIPS NEAR SIRIUS POINT WHICH DROPPED NUMEROUS DEPTH CHARGES X EVADED THOSE VESSELS IN HOPES CONTACTING CONVOY X NOW BELIEVE VESSELS WERE MERELY ASSIGNED PROTECTION

HARBOR MODERATELY HEAVY GUNFIRE TO NORTH OF KISKA X
HAVE TEN TORPEDOES FORWARD REMAINING X FROM UNIT EIGHT
FIVE SIXTEEN.

On August 1, *Grunion* was radioed to continue her voyage to Dutch Harbor, and
in a twist of fate the submarine was to expect to reload and re-provision the boat
and be back at sea on August 3. There was no confirmation of that message. All
further efforts to communicate with *Grunion* were unsuccessful.

With the realization that *Grunion* would never reach port again, the submarine was decommissioned, on paper, on October 5, 1942, and listed as "Missing.
Cause Unknown."

Jim Abele was awarded the Navy's highest honor, the Navy Cross, for his actions
in the Aleutians. Abele's widow, Kay, felt the award belonged to all of the men of
Grunion and acquired the addresses of every man's next of kin. She wrote each
family a letter sharing that the Navy Cross belonged to the submarine's entire crew.
She received many letters in return, a correspondence that would prove valuable
more than sixty years later.

After the war, Japanese records were combed for any clues of *Grunion*'s fate, but
nothing was forthcoming. The submarine remained on the record as missing, cause
unknown, for sixty-five years.

THROUGH THE GENEROSITY OF OTHERS

The sea does not give up its secrets until it is ready, and in the years following World
War II, a few clues to *Grunion*'s activities came to light, but they would go unnoticed
until a diverse group of people crossed paths.

In 1963, Seiichi Aiura, a Japanese freighter captain who was on *Kano Maru* off
Kiska Island at the end of July 1942, wrote an article about an encounter with a
submarine in a seafarer's journal titled *Maru*. Written in Japanese, the article was
printed, then archived, and soon faded from memory.

Thirty-five years after Aiura's story was published, in 1998, retired Air Force Lt. Col.
Richard Lane was browsing in an antique store in Denver, Colorado. In a pile of World
War II memorabilia, Lane found a wiring diagram for a deck winch. The diagram was
covered in Japanese characters and was marked for a ship named *Kano Maru*. Lane
paid one dollar for the intricately marked sheet, took it home, and filed it away.

Two pieces necessary to solve the whereabouts of *Grunion* were now in place.
They would be important pieces in bringing a number of people together who
could, collectively, solve the mystery.

In the years after the end of the war, life was busy for the Abele family, yet they often
wondered what happened to Jim, his crew, and his submarine. With reverence for
their father and his men, the brothers built their lives. The oldest of the three, Bruce,
went into computers; middle son Brad (1933–2008) graduated from Yale University,
became a naval aviator, and flew A-4 Skyhawks during his career; and youngest son
John cofounded the medical device manufacturing firm Boston Scientific.

After retiring, Brad began writing a history of his father titled, simply, *Jim*. The
Jim Book, as the family calls it, was intended for distribution to descendants and
friends, and no one ever expected it to go any farther. While writing the book, Brad
made contact with many of the men who had served with Jim Abele, and the book

The Japanese freighter *Kano Maru*, seen here beached on Kiska Island, was *Grunion's* target that fateful day. *Grunion* fired six torpedoes, one of which exploded in *Kano Maru's* engine room, causing the list visible in this photo. *Richard Lane Collection*

would serve, at that time, as a chronology of what was known of *Grunion's* loss and how Kay Abele handled a difficult living and financial situation in the years following the sub's disappearance.

In 2001, Richard Lane became curious about *Kano Maru* wiring diagram and searched the Internet for information about the ship. Unable to find anything, he posted scans of the diagram on the Japanese military history website J-aircraft .com, not expecting to ever hear anything about the obscure ship. The following day, the J-aircraft.com message board had a post from a gentleman living in Japan named Yutaka Iwasaki. Along with information on *Kano Maru*, Iwasaki attached his translation of the obscure, first-person story written by Seiichi Aiura detailing *Kano Maru's* encounter with a submarine that appeared to be *Grunion*. Lane contacted Darrell Ames, public affairs officer at Commander, Submarines, Pacific (COMSUBPAC), who felt it important enough to post the information on their *Grunion* website.

Bruce's oldest son is named Kurt. His fiancée at the time, Alicia, worked with a gentleman named Ronald Vartanyan, a World War II history buff. Alicia shared the *Jim Book* with him, and Vartanyan sent Bruce a number of *Grunion* websites for review. Bruce followed his advice and at that point discovered COMSUBPAC's post sharing Iwasaki's information about the loss of *Grunion*. Needless to say, because it was the first clue to the loss, it created immense excitement in the Abele family and started an extensive search for contact information with Iwasaki.

Searching the web, John Abele finally found Iwasaki's email address. He sent a note asking, "Are you the one who knows something about *Grunion*?" He received a note back saying, "It's me, I pray for the repose of your father's soul." The one-dollar blueprint found in an antique store led the brothers to a man in Japan who would continue to play a pivotal role in locating *Grunion*.

Several years later while attending a medical conference in Florida, John attended a presentation by Bob Ballard, the man who led the teams that discovered *Titanic*, *Bismarck*, and numerous other historic shipwrecks. Ballard and John got to talking about the search for *Grunion*, and there was the thought of hiring the famous explorer to aid in locating the submarine.

Schedules did not work out as Ballard was booked for the next season, but he did give the brothers a "Submarine Search 101" class. Ballard educated the Abeles on using side-scan sonar for the initial search and examining the most promising contacts with a remotely operated vehicle (ROV), as well as a frank discussion about the strengths and limitations of each technology. Ballard told the brothers they could expect to cover two hundred square miles of sea bottom in about two weeks' time—if the weather was good—and that rocks and other geological formations could distort sonar images and even hide entire ships. Repeating his proven search method used on *Titanic* and *Bismarck*, Ballard told the brothers to "follow the debris trail." In addition, Ballard encouraged the brothers to develop a better picture of where exactly the sub might rest, in order to improve their odds at finding it.

After Ballard's visit, the brothers were more enthused about the potential of a search, but there were still huge obstacles ahead of them. The brothers needed to address issues such as what type of vessel would be best for the search in the rough waters around Kiska, where does one rent a side-scan sonar capable of doing the job, and where on the surface of the ocean to start the search.

It *is* a small world, and after what started as a chance encounter with Bob Ballard, Bruce's wife, Susan, was a dinner guest at her neighbor Arlene Lowney's home. That evening the two ladies were discussing the *Grunion* search and its difficulties. Lowney mentioned that her son Pete had recently returned from fishing the crabbing grounds in the Bering Sea. Susan put Bruce in touch with Pete, and shortly afterward

During the search for the *Grunion*, Yutaka Iwasaki, center, flanked by John Abele on the left and Bruce Abele on the right, was instrumental in many aspects, most notably finding the documents that narrowed the search area from two hundred square miles to virtually pinpointing the exact location. *Courtesy of the Abele family*

the two men had a long discussion about the Aleutian Islands and the hazards of working and sailing there. During the conversation, Pete Lowney referred Bruce to a fishing boat captain named Kale Garcia, owner of the crab boat, *Aquila,* that worked that area in the Aleutians.

Aquila is a 165-foot-long vessel with its wheelhouse forward, leaving a large, open deck that covers the entire aft section of the ship. *Aquila* had been built to service the oil exploration industry, then converted to a fishing vessel. Two cranes, one on each side of the deck, are used for handling crab pots and other fishing gear, and they would be ideal for lifting, launching, and recovering the side-scan sonar sled and ROV. *Aquila* was crewed by Garcia's wife, Anji, and their two teenagers, daughter Kensie and son Tanner.

"I called Kale Garcia, and it just so happened that there was an opening in the boat's schedule and they could handle a trip in August. He was very willing," Bruce said. "I don't know if you know what it costs to rent a boat to go out to Kiska, but it's beyond comprehension. Kale Garcia was absolutely fascinated with the mission and offered to do it for a very, very reduced figure, which made a tremendous difference. On the other side of the coin, we still had to get a side-scan sonar company to assist us, and Garcia named a company he thought might be able to help."

Bruce phoned the sonar company referral, but their equipment could not go deeper than one thousand feet. They in turn referred him to Williamson and Associates in Seattle, where, coincidentally, *Aquila* was berthed. A bit leery of the job of looking for a lost World War II submarine proposed by a man on the phone, Art Wright from Williamson and Associates went to see if *Aquila* would be

Aquila is owned by Anji and Kale Garcia, and their teenagers, daughter Kensie and son Tanner, are valued members of the crew. Kale's interest in helping to search for *Grunion* enabled the submarine to be found at long last. *Courtesy of the Abele family*

Aquila, 165 feet long, was originally constructed as an offshore oil-industry support vessel. It was later converted for the fishing industry, and its two cranes, one on each side of the deck, were ideally suited for sonar tow fish and ROV operations. *Courtesy of the Abele family*

appropriate for the job. It was a little rough, but it looked like it would work. John Abele happened to be in the area at the same time, and a contract was consummated.

But before an expedition could be mounted, the Abele brothers heeded Bob Ballard's advice—narrow the search area as much as possible before leaving the dock. At this point, the search area was two hundred square miles of water off the north end of Kiska Island, one of the most inhospitable places on the face of the Earth.

The Abele brothers needed a more precise location for their dad's submarine. It was Yutaka Iwasaki who came to the rescue. Iwasaki knew a gentleman named Minoru Kamada, who had access to the archives at Japan's National Institute of Defense Studies. Iwasaki traveled by train overnight to meet Kamada to visit the institute, and the pair arrived at the next morning at 6:00 a.m., only to find it did not open until 9:00. Once inside, the archivist informed the weary travelers that the institute did not hold any information about noncombat vessels. "Undeterred, they did a little looking anyway, and in about five minutes they found a pile of two hundred and seventy documents, all about *Kano Maru*," said Bruce. "There they found a document that gave us almost the exact location of where the confrontation had occurred!"

A story in a magazine. A wiring diagram for an unknown ship under a pile of stuff in a Colorado antique store. A chance meeting with the world's foremost ship-wreck explorer. A dinner conversation, a late-night talk with a fishing boat captain, a boat with time to fill in its schedule, a sonar search company that took a chance, and two tired guys not taking no for an answer at an archive halfway around the world. All combined to present the approximate search area for *Grunion*.

ON THE HUNT

Williamson and Associates had to manage the logistics of getting everything required for the search up to Dutch Harbor and on to *Aquila*. In Seattle, a barge was loaded with two side-scan sonars, two very large winches, several miles of cable, and two forty-foot containers that would serve as the sonar shack and the search headquarters. The barge was towed up to Dutch Harbor and its contents transferred and secured on *Aquila*. It was only while the barge was en route that the team learned of Iwasaki's discovery, which greatly narrowed the search area. The 2006 search effort was underway.

The Kiska area is volcanic. The indications were that the sub was on the side of an "extinct volcano that sloped away to the north with deep ridges almost the height of a reasonable-size office building," said Bruce. "Now the problem was they were dragging the side-scan behind the boat and it had to be lowered down into the valleys. If you saw a target, you had to let the side-scan sled move horizontally; up-and-down movement would give unusable images, so it had to be held steady and then reeled in very rapidly so that it wouldn't hit the ridge on the other side. If it did, a three-million-dollar sonar would be lost.

Considering the plans for the 2006 search, the Abeles felt that it was appropriate to attempt to notify the relatives of the crew. Bruce's wife, Susan, while surfing the web, came across a woman named Rhonda Ray. Bruce wrote Ray a note saying: "This is a stab in the dark. I am under the impression that you are a relative of somebody who was on *Grunion*. This August the three sons of Mannert Abele are sponsoring a search for *Grunion* based on new evidence that has been provided because of a Japanese history buff. It is quite a story." Ray wrote back: "Well your 'stab' hit home."

In 2006, *Grunion* was located on sonar using this tow fish. With a short operational window, a trip the following year was made to locate and visually identify the wreck using an ROV. *Courtesy of the Abele family*

From there Ray located Mary Bentz and Vickie Rodgers, who also had relatives onboard *Grunion*, and the amateur genealogist trio began searching for other relatives of *Grunion* crew. Using Kay Abele's postwar correspondence with the next of kin as a starting point, the women found relatives for every crewmember—not an easy task considering that names and locations had changed over the years.

"At two a.m., early in the morning of August 15, 2006, John and I received an email from Art Wright, the project leader from Williamson and Associates, saying that they were ninety-five percent certain that they had found *Grunion*," said Bruce. "It was just unbelievable! The target was about twelve miles north of Kiska almost exactly where predicted by Iwasaki and Kamada's search. It appeared to have lost a significant amount of the bow and to have slid two-thirds of a mile down the side of a dormant volcano."

The next year, 2007, John led a second trip to the area, this time with an ROV equipped with HD video cameras, the goal being to determine whether what had been spotted the previous year was *Grunion*. To operate the ROV, the team needed relatively calm waters, else they could break the cable attached and the multimillion-dollar ROV would be lost. "Their first night in the search area they were expecting a storm to blow in," said John. "They lowered the ROV into the water for an all-night search, and within twenty minutes after the ROV got down to the bottom, they spotted the sub."

Then the ROV took a wrong turn and they lost contact with *Grunion*. It took another hour and a half to get back onto the target, which finally happened when they came across the slide path (exactly as Ballard had advised: "follow the debris path"). One of the first anomalies the team noticed was that the aft battery hatch was wide open. This led to early speculation that the crew was making a surface attack on *Kano Maru*.

"The crew of *Kano Maru* reported that the submarine fired six torpedoes," said Bruce. "The first one missed. The second one hit the engine room and disabled *Kano Maru*. The third one was bad, went under the freighter, and didn't explode. The next two hit *Kano Maru* but bounced off, neither one exploding. The last torpedo appeared to have been a circular run, not uncommon in those days, that came around and struck the submarine in the periscope shears, but it didn't explode."

The evidence suggests that the sonar man heard the high-pitched whirring of the torpedo as it changed course and the skipper ordered a hard dive to get out of the torpedo's way. Traveling at fifty-three miles per hour and weighing 3,500 pounds, a torpedo has the same impact velocity as a small car, so it can do a lot of damage—even if it doesn't explode. With the hard dive order, the dive planes were put into a downward position, but the submarine was still not able to get out of the torpedo's way. It was struck in the periscope shears, the structure holding the periscopes, where a significant dent is plainly visible. The torpedo broke both periscopes, in essence blinding the sub, but did not explode.

In analyzing the video of the wreck, it was noticed that the rear dive planes were still in the dive position. The analysts

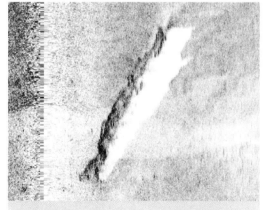

Sonar image of what turned out to be *Grunion*. It doesn't look like a sub, but it has the length and was certainly at the top of the list of targets to inspect with the ROV. *Courtesy of the Abele family*

The *Grunion* search team on board *Aquila*. Front row, left to right: Tanner Garcia, Kensie Garcia, Ian Walt, Gromeko Lekka. Second row, left to right: Peter Lowney, Anji Garcia, Kale Garcia, John Abele, Rich Abele, Rich Keeling, Dave Gallo, Abby Fammartino. Back row, left to right: Toshi Mikagawa, Richard Graham, Joe Caba, Donavan Webster, Mike Nicholson. *Courtesy of the Abele family*

surmised that during the descent, the dive planes became jammed, forcing the submarine down at a steep angle from which it was unable to recover. The sub plunged down until it reached about one thousand feet, where it imploded, instantly killing the crew and blowing one hatch open.

When the sub impacted the bottom, fifty-two feet of the bow broke off; the remainder then slid two-thirds of a mile down on the side of an extinct volcano, coming to rest in a notch in the terrain.

Seventy men perished at this spot; the families now knew where it happened and why.

GRUNION'S BELL STILL TOLLS

To publicly recognize the sacrifices made by the *Grunion* crew, a memorial was planned for October 11, 2008, at the USS *Cod* Submarine Memorial in Cleveland. *Cod* is a sister ship to *Grunion*. The "Sub Ladies," as Ray, Bentz, and Rodgers had become known, invited all of the crewmembers' next of kin to the ceremony. About half of the seventy crewmen were represented at the memorial by a relative, all located through the Sub Ladies' efforts.

For the memorial, it is traditional to remember the fifty-two lost US submarines in a ceremony called the "tolling of the boat." A bell is struck once as each submarine's name is read. It is a very solemn ceremony, and although *Cod* (SS-224) has its own bell, many thought it would be fitting if members of *Grunion* could be present at the ceremony.

As it turns out, one of *Grunion*'s crew was from Greenville, Mississippi. Stan Kendrick and his wife, Geraldine, sister to Seaman 2nd Class Edward Knowles, had sent Bruce Abele a photo of the sub's bell on display in the town's visitors center.

Before leaving Pearl Harbor for combat in the Aleutians, the crew of *Grunion* removed anything from the boat that might make noise or come loose under water and put those items into storage. The bell was an obvious choice.

The glass faces of *Grunion*'s range finder, located on the conning tower, were blown off by the water pressure. *Courtesy of the Abele family*

Close-up of the periscope, which, like the range finder, has been blown out by the water pressure. The periscope may have been damaged or distorted when the circularly running torpedo struck the periscope shears. *Courtesy of the Abele family*

"During the Korean War, a Navy chaplain named Noonan was stationed in Pearl Harbor, saw the bell that had been removed from the ship in 1942, and asked if he could have it to take back to Mississippi," Bruce said. "The Navy said no, yet a few months later, while Noonan was in the South Pacific, a hundred-pound package showed up at his base. Lo and behold, it was the bell. When he retired, Noonan moved back to Greenville, Mississippi. He was ninety-eight when he passed away, and he willed the bell to the city, who placed it on display."

The first photos of the submarine were of its stern. Notice the rudder, diving planes, propellers, and (at the top of the photo) the propeller guards to prevent the sub from striking the tender or the dock. The laser pointer is focused on the diving planes, which are in the full-down position. *Courtesy of the Abele family*

The after battery room hatch was blown fully open when the submarine imploded. At first, many surmised that the crew had tried to escape the submarine, but upon closer inspection the hatch dogs were determined to have sheared off from the pressure. *Courtesy of the Abele family*

After obtaining permission from the US Navy, the Abele brothers were able to borrow *Grunion*'s original bell and had it shipped and on hand for the ceremony.

There was another surprise forthcoming. Shortly after the memorial, Bruce was contacted by Takuya Asakura, a reporter who had done several excellent stories on *Grunion* for Japan's largest newspaper, *Asahi Shimbun*. Asakura had received a note from Kazuo Shinoda, the son of the commander of one of the sub chasers sunk by *Grunion*. His mother, Chiyo, ninety-eight at the time, said that just before her husband, Isamu, was killed, she received a note from him with flowers from Kiska, which she still had carefully preserved all these years. Seventy years later, the Abeles sent her pressed flowers from Kiska from the "Sons of Catherine E. Abele." It made the front page of several international papers.

An interesting aside to this story occurred in June 2014. Bruce received an email from Michael Mohl, an Israeli who runs the website NavSource. He had received a note from a Japanese farmer named Yobu saying that he had found a document indicating the

This distorted area of the hull is where the implosion occurred. *Courtesy of the Abele family*

SPECIFICATIONS USS *GRUNION* (SS-216)

Length	311 feet 9 inches
Beam	27 feet 3 inches
Draft	16 feet 10 inches
Displacement	1,525 tons surfaced; 2,415 tons submerged
Powerplant	4 General Motors 16-248 V-16 diesel engines 4 General Electric electric motors 2 126-cell Sargo batteries
Horsepower	5,400 shaft surfaced; 2,740 shaft submerged
Top speed	20.5 knots surfaced; 8.75 knots submerged
Crew	70
Armament	6 bow, 4 stern torpedo tubes 24 21-inch torpedoes 1 3-inch/50-caliber deck gun
Launched	December 22,1941
Commissioned	April 11, 1942
Decommissioned	October 5, 1942
Builder	Electric Boat Co.
Class	*Gato*
Website	www.ussgrunion.com

location of the wreck of the sub *Escolar* (SS-294—lost October 17, 1944, with eighty-two crewmen on board). What was needed was a translation, and Bruce turned to Yutaka Iwasaki. That document ultimately provided a precise location for that sub in waters only about 450 feet deep and 50 or so miles from land. That is quite different from *Grunion*'s three-thousand-foot depth and location 1,500 miles from the nearest civilization. Tony Duda, a relative of one of that crew, is now coordinating the project, looking for a way to confirm the location.

It is remarkable what collaboration can accomplish.

Grunion sailed from Hawaii to the Aleutian Islands on June 29, 1942, not to be seen for another sixty-five years. *US Navy*

I-400 CLASS

JAPAN'S PANAMA
CANAL KILLERS

JAPAN'S *I-400*-CLASS *SENSUIKAN-TOKU* (special attack, typically shortened to *Sentoku*), aircraft-carrying submarines, were a weapon ahead of their time, and one that could have had an impact on the war had they been deployed when the war's opening salvos were fired. In hindsight, these submarines' destructive capabilities could have been used in a number of catastrophic attack scenarios, such as dropping nuclear, chemical, or biological weapons on American cities, but thankfully they were not.

Although there had been aircraft-carrying submarines prior to the Sentoku boats, *I-400*-class special attack subs were unlike anything ever seen before. Admiral Yamamoto ordered eighteen of the type in 1942. They were huge by any standard, could hold enough fuel that they could cruise on the surface at fourteen knots, and had a range of 37,500 miles—one and a half times around the world. Each Sentoku submarine carried three Aichi M6A *Seiran* floatplanes, capable of delivering either one Type 91 aerial torpedo (fitted with a 518-pound warhead) or one 1,874-pound bomb.

The Sentoku submarines were approximately 400 feet long, with a beam of 39.3 feet and a draft of 23 feet. The subs displaced 5,700 tons submerged and carried a crew of twenty-one officers and 170 sailors. *I-400* class statistics show that each of these submarines was longer than the US Navy's most numerous destroyer type, the 376-foot-long *Fletcher* class, of which 175 were built, and they were 100 feet longer than the majority of American World War II fleet submarines. It would take the US Navy until August 19, 1958, to eclipse the length of the Sentoku submarines, when the radar picket boat *Triton* (SSRN-586) was launched.

The three Seiran aircraft (a name translating roughly to "Clear-Sky Storm") were carried in a centerline hangar, stowed with their wings folded and without landing gear. The centerline hangar required that the conning tower be offset; designers placed it to port of the centerline, making the conning tower more than thirty feet tall when measured from the submarine's deck.

Floats for the fighters were carried in under-deck compartments and installed prior to launch should the mission profile call for the fighter to return to the ship to be reused. In attack scenarios, the Seirans would be launched without landing gear or floats, and upon return the fighters would be ditched and the pilots picked up by the submarine and subsequently returned to duty.

Although of poor quality, this is the only known photo of M6A Seiran aircraft on the launch ramp of a submarine. Judging from the length of the aircraft hangar, this is most likely *I-14*. The wings on both Seirans are folded and neither has floats installed. *John Geoghegan Collection*

The first of the Sentoku submarines, *I-400*, was laid down on January 18, 1943, at the Kure Naval Arsenal, with *I-401* following on April 26 and *I-402* on October 20, the latter two both at the Sasebo Naval Arsenal.

The sole surviving intact M6A Seiran was sent to the Naval Air Station at Alameda, California, after the war. It is seen here at the 1960 base open house. This and a number of other historic aircraft were maintained in a small on-base collection and brought out on special occasions. This aircraft was later donated to the National Air and Space Museum and today can be seen, fully restored, at the museum's Udvar-Hazy Center at Dulles Airport in Chantilly, Virginia. *US Navy*

Due to a shortage of steel needed to construct the giant submarines, *I-403* and *I-406* through *I-417* were canceled before winter 1943. *I-404* was heavily damaged during a bombing raid on Kure on July 28, and *I-405* was scrapped on the building way. *I-400* was launched on January 18, 1944, and completed on December 30 of that year. *I-401* was launched on March 11, 1944, and completed on January 8, 1945, and *I-402* followed into the water on September 5, 1944. *I-402* was subsequently finished as a submarine tanker to carry fuel oils to beleaguered Japanese garrisons on the islands in the Pacific Ocean.

Initially, the aircraft-carrying *I-400* submarines were intended to attack New York and Washington, D.C., but by early 1944, Japan's changing fortunes of war and the Allies' greatly improved antisubmarine warfare capabilities eliminated the US East Coast from the list of potential targets. Japanese planners then changed their focus to attacking the Panama Canal. They estimated an attack on the canal would prevent its use for at least six months, wreaking havoc with the Allies' ability to supply and reinforce the Pacific Fleet.

"*I-400* submarines were way ahead of their time. What makes them so significant is that no sub had ever been employed in an offensive manner before *I-400* subs,"

Japan's three largest aircraft-carrying submarines are seen at Pearl Harbor in early 1946. From left to right, *I-400*, *I-14*, and *I-401* are each moored with their aircraft hangar doors open. Note the offset conning towers and the antiaircraft guns on the platforms above the hangars. The submarines' inclined aircraft catapult launch ramps can also be clearly seen in this photo. To the left, across the channel, American submarines can be seen docked at the sub base. *Everett Leavins, courtesy Auburn University Libraries, Special Collections and Archives*

said John Geoghegan, *I-400* class submarine expert and author of *Operation Storm: Japan's Top Secret Submarines and Its Plan to Change the Course of World War II.* "These were the first subs ever designed to attack cities and, of course, the whole basis behind America's nuclear deterrent strategy is to use submarines as stealth weapons to launch nuclear missiles against our enemy's land-based targets."

Japan's *I-400* class were important enough that after capitulation they were sailed to Pearl Harbor for a complete intelligence assessment, then destroyed before they could be studied by the Soviet Union. Lessons learned from *I-400*-class boats took a decade to incorporate, and by the mid-1950s the US Navy began deploying submarines with hangars capable of launching nuclear warhead-equipped SSM-N-8 Regulus missiles. "Our ballistic missiles boats, the boomers of today, are the great, great grandchildren of *I-400* subs," said Geoghegan. "Until the development of *I-400* class, submarines were focused on scouting, or sinking merchant and capital ships. There wasn't a serious, submarine-borne, land-target attack capability until *I-400* and *I-401.*"

By summer 1945, fuel was in extremely short supply, and the mission to bomb the Panama Canal was shelved. Both submarines sailed for Manchuria to take on fuel for their first mission—to bomb the Allied anchorage at Ulithi Atoll, in the Caroline Islands. Ulithi Atoll is 1,300 miles from Tokyo, and its protected anchorage is twenty miles long to the north and south and ten miles wide from east to west. At its peak in June 1945, Ulithi Atoll held 722 ships in preparation for the invasion of Okinawa; thus, the atoll was a prime target for the Japanese.

For the Ulithi Atoll attack, *I-400*-class submarines were slated to surface and launch their Seiran fighter-bombers on August 17, 1945. However, hostilities were halted on August 16, and the submarines were ordered to turn back and sail to Japan.

CAPTURE AND DESTRUCTION

On August 27, 1945, at a position 250 miles northeast of Tokyo off the eastern coast of Japan's Honshu Island, aircraft from the US Navy's Third Fleet Fast Carrier Task Group 38.1, under the command of Vice Adm. John S. McCain, spotted a surfaced submarine flying a Japanese flag and a black flag of surrender. An hour later, a second surfaced submarine was spotted cruising at ten knots in the same general vicinity.

Destroyers *Murray* (DD-576) and *Dashiell* (DD-659) were assigned to locate and board the submarines. The first vessel to be encountered was *I-14*, a 376-foot-long aircraft-carrying submarine. Although twenty-five feet shorter than *I-400*-class boats, *I-14*'s hangar had room for two M6A1 Seiran fighters and was nearly equal in length to the destroyers that captured her. A boarding party from destroyer *Murray* accepted the Japanese submarine's surrender at sea, and then escorted the war prize to Sagami Wan, Honshu. A prize crew from the submarine tender *Proteus* (AS-19) was sent out to meet *I-14* and her US Navy escorts. At the rendezvous point, the American sailors were put onboard and an equal number of Japanese submarine crewmen were transferred to destroyer *Bangust* (DE-739). *Bangust* relieved *Murray* and escorted *I-14* to Sagami Wan, where she tied up to *Proteus.*

Later on the afternoon of August 27, destroyers *Blue* (DD-744) and *Mansfield* (DD-728) located the second submarine some five hundred miles northeast of

Tokyo. This submarine turned out to be *I-400*. This sub was also escorted to Sagami Wan and was tied up to *Proteus*.

On August 29, at nearly 11:00 p.m. local time, the US Navy submarine *Segundo* (SS-398) acquired *I-401* on its radar. After an all-night pursuit, *Segundo* made contact with *I-401* and ordered her to stop. The Japanese sub's navigation officer, Lt. Bando Muneo, was sent to *Segundo* to give details of the vessel and its intentions. Upon his return, Muneo reported the Americans' intentions to sail the boat to Yokosuka, but *I-401*'s captain objected and considered scuttling the boat. The US Navy crew chained open the sub's hatches to prevent it from submerging, and the captain and other officers were removed to the American destroyers. Now under American control, *I-401* sailed for Sagami Wan, the bay southwest of Yokosuka (with Yokosuka located over the peninsula on Tokyo Bay).

All of the aircraft-carrying submarines were present in Tokyo Bay on September 2, 1945, for the formal surrender of the Japanese Empire. Two months later, US Navy prize crews began the task of sailing or towing *I-201, I-202, I-14, I-400*, and *I-401* to the submarine base at Pearl Harbor to assess the intelligence value of the Japanese vessels, with *I-14, I-400*, and *I-401* departing Japanese waters on November 1, 1945.

Everett Leavins was an Army Air Forces aircraft mechanic stationed at Luke Field, on Ford Island in Pearl Harbor. Leavins arrived in Hawaii in 1931 and, as an avid photography buff, captured many of the events surrounding military life on Oahu. When *I-14, I-400*, and *I-401* arrived at Pearl Harbor, Leavins captured the submarines on film, taking a number of interesting detail photographs. The size of the submarine and its aircraft hangar are quickly visible when compared to the officer about to cross the gangway in the right of the photo. Also note the size of the hangar door, and the length of the hangar itself. *Everett Leavins, courtesy Auburn University Libraries, Special Collections and Archives*

One of the most interesting aspects of *I-400*-class submarines was that the hulls were riveted, rather than welded like American fleet boats and German U-boats. *I-400* boats' riveted hulls limited their dive depth to 330 feet, and it is estimated that during a severe depth charging the rivets would quickly buckle and break. The loss of rivets would lead to water incursion, and if enough of the hull was breached and the sub was heavy enough, it would quickly descend, exceed its maximum crush depth, and implode. *I-400*-class submarine crews' chance of surviving a prolonged depth charging attack was minimal at best.

Prior to surrendering, *I-14, I-400,* and *I-401* had scuttled their Seiran aircraft, jettisoned their torpedoes, and tossed their codebooks and sensitive papers overboard in weighted bags, leaving behind vessels of questionable intelligence value. Both *I-400*-class vessels had eight torpedo tubes, yet the massive submarines carried only twenty torpedoes—a small load for a submarine that could, potentially, be at sea for more than six months and far away from any potential replenishment.

I-400-class submarines' offensive capability was based primarily around their aircraft. And although they were not onboard the captured submarines, what was of intelligence value was the aircraft hangar, its watertight door, and the infrastructure used to support Seiran operations while at sea.

To prevent the Soviet Union from having the opportunity to evaluate captured Japanese submarine technology, the US Navy disposed of the five captured vessels in short order. Beginning in May 1946, a series of live-fire torpedo exercises was

SPECIFICATIONS: *I-400* AND *I-401*	
Length	400 feet
Beam	39 feet 4 inches
Draft	23 feet 0 inches
Displacement	6,670 tons
Powerplant	4 diesel engines producing 2,250 shaft hp each; 2 1,600-kilowatt electric motors; twin screws
Speed	18.7 knots surface; 12 knots submerged
Maximum depth	330 feet
Crew	144 officers and sailors
Armament	8 21-inch bow torpedo tubes 1 140-millimeter deck gun 3 25-millimeter machine guns (triple mount) 1 25-millimeter machine gun 3 Aichi M6A1 Serian floatplane fighter-bombers
Builder	Kure (*I-400*), Sasebo (*I-401*)
Launched	January 18, 1944 (*I-400*); March 11, 1944 (*I-401*)
Commissioned	December 30, 1944 (*I-400*); January 8, 1945 (*I-401*)
Sunk	*I-400* sunk by *Trumpetfish* (SS-425), June 4, 1946; *I-401* sunk by *Cabezon* May 31, 1946

held in the waters off Barbers Point, Oahu, Hawaii. On May 21, *I-203* was the first to go to the bottom when *Caiman* (SS-323) sent a Mk 18, Mod. 2 torpedo into her side. Two days later, *I-201* followed her to the bottom, sunk by *Queenfish* (SS-393). On May 28, torpedoes from *Bugara* (SS-331) testing the new Mk 10, Mod. 3 warhead exploder split *I-14* in half, and on the last day of the month, *I-401* met her end when *Cabezon* (SS-334) took aim and fired. Four days later, on June 4, *Trumpetfish* (SS-425) sent a trio of Mk 18, Mod. 2 electric torpedoes, which were testing the Mk 10, Mod. 3 exploder, into *I-400*'s side. At the time, everyone thought the world's largest submarines were gone forever.

FINDING THE SUNKEN SENTOKU SUBMARINES

The majority of the historic World War II vessels discovered in Hawaiian waters within the past thirty years involved the University of Hawaii's *Pisces* submersibles. Operated by the university's Hawaii Undersea Research Laboratory, known as HURL, Terry Kerby serves as the director of facilities and submersible operations. Kerby served in the US Coast Guard as a quartermaster and navigator on ships based in Alaska, at various ports in the Gulf of Mexico and the Caribbean Sea, and on the US East Coast. After serving, Kerby trained as a commercial diver and began his association with submarines in 1976 while working for Maui Divers on the submersible *Star II*.

HURL operates two submersibles, *Pisces IV* and *Pisces V*, for traditional oceanographic and Earth sciences research. Each diving season, the submersibles are required to make three test dives, and these descents are used to verify the performance of the sub's systems at depth. On these test dives, shipwrecks provide excellent targets to validate performance of the sub's sonar, manipulator arms, cameras, and other tools vital for undersea scientific research. Once the three test dives are completed, the mini submarines depart on three- to five-month deployments to areas of interest around the Pacific Ocean.

While making test dives each season, HURL's Steve Price began assembling a database of historical objects on the ocean floor south of Oahu. Price has recorded every aircraft, pile of ammunition, automobile (some going back to 1915), sailing ship, modern shipwreck, and mass dumping site. "Steve has an incredible database," said Terry Kerby. "When we do our test and trial dives, we build upon that database every year. No agency has ever funded dives to go out and do this work; it's all been part of our yearly preseason testing and training. Once we find a target of significant historical value, then the National Oceanic and Atmospheric Administration's National Marine Sanctuaries, the National Park Service, and/or film documentary groups have funded dives to continue the investigations. It has really paid off over the years. There are groups again this year [2014] that paid to go back to *Ward*'s midget sub. That has been happening continuously since 2002."

By chance, the first Sentoku submarine located by the HURL submersibles was *I-401*. When *Cabezon* sent torpedoes into its side, *I-401* settled more than 2,690 feet below the surface. Kerby said:

> We thought that they wouldn't be that difficult to find because there were positions from the Navy subs that sank them off of Barbers Point, but not a

The ship number painted on the conning tower of *I-401* is still visible after sixty-nine years underwater. The submarine was sunk off the southern coast of Oahu by torpedoes from the US Navy submarine *Cabezon* (SS-334) on May 31, 1946. *HURL*

single one of those subs was anywhere close. Some of them were as much as eight miles away from their original recorded positions. It took us a while to find them. We couldn't look every year, but in 2005 we had a really good test dive. We were preparing to go from Hawaii to Samoa and then start on the Samoa hot spot volcano by Vailulu'u. For five months, nonstop, we dove on thirteen active submarine volcanoes between Samoa and New Zealand and back.

On our first test dive, we found the wreck of the *S-19*, an old World War I S-boat that was sunk on December 18, 1938, to comply with the Second London Naval Treaty that everybody adhered to—except for Japan. Then, on our second day, we were in the general area where we thought that some of those I-boats might have ended up.

On our dives, we deploy both the *Pisces IV* and *Pisces V*. We do emergency tracking exercises where one sub runs off and hides and each one of the pilots has to put on the headphones and try and home in on the other sub. It's part of our exercises. I was in *Pisces IV* with one of my other pilots, Colin Wollerman. The other pilot, Maximilian "Max" Cremer, was in *Pisces V* with Steve Price. Wollerman and I were in an area where there were some rock formations, and we thought we were coming up on another big one. Out of the dark comes the broken-off end of the bow section of the *I-401*. It was like we were looking at something from a three-story-tall building. It was all torn up.

We maneuvered down to the bow and called *Pisces V* to say we found an aircraft carrier submarine. There was a catapult launch ramp, but we didn't know which submarine it was. *Pisces V* was maneuvering to meet up with us, and Cremer called me and they were about 250 meters away to the north. Cremer

I-400 and *I-401* each carried a 140-millimeter Type 11 cannon aft of the conning tower. It was the largest deck gun on a submarine. The wooden deck has mostly rotted away from around the gun, exposing the metal support structure. *HURL*

then radioed, "We found another one." I said, "I think what you found is the main hull that goes with this bow section." Then we started moving up toward where they were.

There was a seven-hundred-and-fifty-foot debris field between the bow section and the main hull. The main hull was sitting upright and the hull numbers could clearly be seen painted on the side. I maneuvered and perched the sub on the edge of the conning tower, and you could look right into the bridge, which is complete. When it tore in half, everything was left intact from just forward of the conning tower back, but the whole hundred-and-fifteen-foot hangar was ripped from the hull. There was a big hole where it used to be.

Pisces V was parked at the stern of *I-400*, some seven hundred and fifty feet up-current of us, waiting for me to arrive. When they settled in, they stirred up a big silt cloud that blew down with the current. Wollerman and I were easing up on some target that rose more than thirty feet off the bottom. It was right in front of us. The next thing, we had absolutely no visibility. We're in the middle of a silt cloud.

Completely blind in a silt cloud could have been disastrous for *Pisces IV*; however, Kerby, having established protocols for working around big wrecks, was able to navigate to safe water. Most submersibles avoid getting close to shipwrecks for fear of becoming entangled in snagged fishing nets or in the wreckage itself. But after having investigated more than twenty-five shipwrecks with hulls more than a hundred feet long, Kerby and his team have established rigid procedures for the initial survey of a wreck. He said:

One of HURL's *Pisces* submersibles hovers over the conning tower of *I-401*. The wreck lies at a depth of 2,690 feet. *HURL*

A close-up of the conning tower with an open hatch in the lower left and the base of the periscope shears to the right. *HURL*

The senior pilot will go in first and do an initial survey while the other sub waits. This way we don't have two subs in the water column having to worry about bumping into each other while conducting the initial safety assessment.

Another thing we do when we approach one of these wrecks is that we stop and draw a compass rose. We draw what we see on our sonar so we get an idea of what is around us and where safe water is. Because we had done that, I knew exactly where to turn to get back to safe water, taking us out of the silt cloud. We eventually went back and discovered that the target we were approaching was the end of the hangar. When it ripped out, it tore part of the hangar apart, and that end of the hangar, with that giant door, was stuck in the sea bottom like the Leaning Tower of Pisa. That's what we were approaching when the silt kicked up. We eventually inspected it years later.

TWO-FER: *I-14* AND *I-201*

When the HURL submersibles find a historic shipwreck, the laboratory typically issues a news release, which is often printed nationally. From the news stories, many of the veterans who served on board these ships will contact HURL with more information, photographs, and sea stories relating to the vessel just found. Kerby said:

> We were hearing from a lot of the sailors that were the prize crews that originally boarded those subs and sailed them back to Hawaii. That's a real reward when we find one of these. One of those guys, named Charles Alger, was chief of the boat of the captured *I-14*. When they reassigned all of the crews, they left the chief of the boat to care for the sub until they took them out and sank them. They let Alger ride out on *Bugara* to witness the sinking of the *I-14*. He shot sixteen-millimeter footage of the sinking and sent us a copy of the film. There it is, 1946 Oahu, and this big, majestic sub setting out to sea. You see the back of the *Bugara,* and then it launches a torpedo. When the torpedo hit, it sent the *I-14* up on her nose and then the hull crashed down. The submarine then disappeared into the boiling water, not to be seen for nearly fifty years.
>
> Alger's film captured 1946 Oahu from Ka'ena Point all the way to Diamond Head. That gave us some landmarks to fix some ranges on. The next science dive season, we used those coordinates to plan our dive to locate the *I-14*.
>
> National Geographic wanted to do a documentary on the *I-14*, but they were really hoping that we would have something new that we could find too. If you ever watch the documentary *The Hunt for the Samurai Subs*, it makes it look like we searched and failed for several dives and then, finally, we hit pay dirt and found the wreck of the *I-14*. Because of Alger's film footage, we were actually

A multitude of sea growth encroaches on the white outlined hull numbers on *I-14*'s conning tower. *HURL*

I-14's hangar is missing its massive door. The hangar is large enough that the *Pisces* research submersibles could be flown inside. *HURL*

The triple twenty-five-millimeter anti-aircraft guns of *I-14* point to the surface. *HURL*

sitting on the bow of that sub less than an hour after we got on the bottom.

The morning we found the wreck of the *I-14*, we found the bow. Then, we followed the debris trail and located a channel that had been made when the main hull section slid down this slope and the rudder carved a groove in the sea bottom. We followed that groove to the main hull, which was about seven hundred and fifty feet away. She was sitting upright as well, with a big *I-14* painted on the side. The hangar was still there, but the front-end hangar door was missing. Sailing up to look inside, the hangar appeared big enough that you could have flown the *Pisces* in there and parked it.

Satisfied with having located *I-14*, Cremer in *Pisces V* ascended for the surface. With time to kill, Kerby and Price in *Pisces IV* turned west to get away from *I-14* before leaving the bottom. Then their sonar presented another big target. "There was another big bow section laying there, like a knife-edged bow. It was the bow section of *I-201*," said Kerby. "The next day, we found the main hull of *I-201*, which is sitting upright on the bottom, again, with the bow about six hundred feet away."

The Japanese had built two fast attack submarines that were launched too late to see service during the war, and *I-201* was one of them. *I-201*-class submarines were 259 feet long, had a beam of 19 feet and a draft of 18 feet, and featured welded steel hulls and anti-sonar/antiradar hull coating. The sub's five-thousand-horsepower electric motors were capable of pushing the boat at submerged speeds up to nineteen knots for periods of about one hour—it was faster underwater than surfaced. The submarine operated with a crew of only thirty-one officers and sailors.

TWO-FER TIMES TWO: *I-400* AND *C. S. DICKENSON/KAILUA*

"We spent a lot of years looking at acres of barren mud," Kerby said about the search for *I-400*. "We basically reverse navigated records from *Trumpetfish*, which put it well off shore and quite a ways out in an area that hadn't been mapped. There wasn't any bottom definitive data that we could find. No shadows or anomalies to go check out. No clues." From the lack of available data, everyone knew it was going to take a long time to find *I-400*.

Using Price's database and other underwater mapping the HURL team had done, they had a number of bottom anomalies that had yet to be investigated. Essentially, a ship uses side-scan sonar to profile the bottom for one scientific reason or another, and the HURL team then looks at the data searching for suitable targets. Upon investigation, the anomalies often turn out to be rocks or other geologic features, but sometimes they get lucky and find historic wreck sites. "In 2014, the National Marine Sanctuaries had funding for three days of maritime heritage dives," Kerby said. He continued:

The super-fast *I-201* lies on its side at the bottom of the Pacific Ocean. The Japanese rising-sun flag, hull number, and chrysanthemum emblem at the base of the periscope shears are painted on the conning tower. Unlike the *I-400-*class submarines, the *I-201-*class boats were streamlined to enable speeds up to nineteen knots while submerged. *HURL*

Usually, Jim Delgado and Hans Van Tilburg, archaeologists with National Marine Sanctuary's program, never risk using their precious funded dives to go look at something that may turn out to be a rock. Over the years we've had a lot of success, but Delgado really took a risk this time. We had two anomalies that we hadn't been able to survey yet. Using one of their previously funded dive days, Delgado chose to check out some of the anomalies rather than returning to known targets like the *Ward*'s midget sub, or the three-piece midget sub, or any number of other wrecks.

We were going to try and squeeze in two dives that day. On the first dive, out of the dark comes the *I-400*. It was much closer to shore than we thought it was. On the second dive, we found the wreck of the *C. S. Dickenson* [later operated by the Navy as the auxiliary ship *Kailua*]. There is an amazing historic photo in Pearl Harbor where someone climbed up on the *Dickenson* and took a picture off the back. The photo shows the *Dickenson*'s teak deck and its big helms wheel. In the photo is the *I-400* tied up behind it. On that day, we found the *I-400* and the *Dickenson* within six miles of each other.

C. S. Dickenson was built by Sun Shipbuilding and Drydock Co., of Chester, Pennsylvania, and was completed in April 1923. The ship was owned by the Commercial Pacific Cable Co. and laid cables between islands in the Pacific. After World War II broke out, it was impressed into service by the US Navy on May 19, 1942, and renamed *Kailua* (IX-71). The ship served in the South Pacific combat zone, traveling as far as New Guinea, before returning to Pearl Harbor in summer 1944. After the war, she was used as a target and sunk on February 7, 1946.

I-401, *I-14*, and *I-400* as seen from the wheelhouse of the US Navy auxiliary ship *Kailua* (IX-71). Terry Kerby and the HURL team found both *I-400*, moored directly behind and against the dock, and *Kailua* on the same day. The first thing Kerby saw was *Kailua's* helm, also seen in the foreground. *Everett Leavins, courtesy Auburn University Libraries, Special Collections and Archives*

"My first view of *Dickenson* was that big helms wheel on the fantail. It was classic," Kerby said. "That ship is totally intact for a ship that was torpedoed. There's no sign of torpedo damage. All the rigging is also up there. It was really a fascinating wreck to explore. Another really historic site to add to our list."

YET TO BE DISCOVERED

With Kerby having found numerous sunken ships and solved a couple of enduring maritime mysteries, what is on his short list of "like to find" historic shipwrecks? At the top is the cruiser *Baltimore* (C-3), which was laid down on May 5, 1887, served in the Spanish-American War and World War I, and was present at Pearl Harbor, serving as a damage-control training ship. A little more than two months after the Japanese attack, on February 16, 1942, *Baltimore* was reportedly sold for scrap; however, many believe the Navy towed the cruiser out of the harbor and scuttled her. Finding *Baltimore* is always in the back of Kerby's mind.

And then there's the Japanese submarine *I-23*. It would be an incredible discovery, and one Kerby would like to make, although he describes it as a "needle in the haystack." *I-23* participated in Operation K, known as the "second attack on Pearl Harbor." Operation K called for a pair of Kawanishi H8K flying boats to sortie from Wotje Atoll in the Marshall Islands, fly 1,900 miles to French Frigate Shoals, refuel from a waiting submarine, then fly south nearly six hundred miles to reconnoiter Pearl Harbor to assess how the US Navy was recovering from the December 7, 1941, attack. Each of the H8K "Emily" aircraft carried four 550-pound bombs that they were supposed to drop on the docks and other targets of opportunity within Pearl Harbor.

For Operation K, *I-23* was deployed to an area south of Pearl Harbor to serve as a lifeguard submarine should either or both of the flying boats be severely damaged while attacking the harbor.

The raid took place on March 4, 1942, but Pearl Harbor was socked in by stormy weather. In addition, the US Army had learned its lesson on December 7, 1941, and was closely monitoring all approaches to Oahu on radar. The Emilys were tracked and Curtiss P-40 fighters were scrambled to intercept them. In the clouds and experiencing radio problems, the two H8Ks became separated, attacking from different directions, but neither could find Pearl Harbor. One of the flying boats dropped its bombs on Oahu's Tantalus Peak, and the other is believed to have jettisoned its bombs into the sea. The P-40s did not locate the H8Ks in the storm clouds and returned to base.

I-23 reported in for the mission on February 14, 1942, and gave reports until the twenty-eighth of the month. The submarine was never heard from again. "We believe that that submarine is still out there," Kerby said. "There's no way they would have abandoned their mission and left. Something happened. There was no report of any antisubmarine warfare during that time. We think they may have had a mechanical failure that caused a loss."

I-23 is 356 feet long, a target size Kerby is very familiar with. If and when he finds it, he'll bring closure to the families of the submarine's ninety-six crewmen and solve another maritime mystery from World War II.

SEEING A PT BOAT IN THE WATER FOR THE FIRST TIME, ONE IS STRUCK BY THE DEEP V OF THE HULL AND HOW THE BOW SITS VERY TALL IN THE WATER. STEPPING ABOARD AT THE BULLNOSE, THE BRONZE SEMI-CIRCLE THAT GUIDES ROPES AND THE ANCHOR CHAIN, VISITORS ARE GREETED BY A LARGE VARIETY OF ARMAMENT. THIS BOAT IS BUILT FOR ONE PURPOSE, AND EVERY INCH OF THE MAIN DECK IS USED FOR OFFENSIVE ARMAMENT.

BACK FROM
THE BRINK

ONCE-DISCARDED PT BOAT NOW FULLY RESTORED

PT BOATS ARE THE STUFF OF LEGENDS. They are known by various names: Motor Torpedo Boat, Patrol Torpedo, PT, "the Mosquito Fleet," or, if you were a Japanese sailor during World War II, "Devil Boats."

The World War II exploits of PT boats and their exemplary crews are dramatic to say the least. The boats and their crews saved a general, harassed the enemy and sank numerous ships, made the reputation of a US president, and earned two motor torpedo boat squadron commanders the Medal of Honor.

Yet, once the war ended, many PT boats were stripped, burned, and abandoned. Those that sailed on after the conflict were tools of war, not to be respected

or cherished but used as a means to complete a task, whether it was patrolling as originally intended or some other utilitarian work that needed a small, fast, and sleek boat.

World War II's PT boats got their start in 1930s England before the conflict. The British Power Boat Company designed a sixty-foot-long motor torpedo boat, of which the British Admiralty bought six for trials in 1934. Four years later, the Vosper Co. floated a sixty-eight-foot motor torpedo boat that could carry two torpedoes. Again, the Admiralty ordered six examples. In 1939, the British Power Boat Co. showed off its new seventy-foot design capable of carrying a pair of twenty-one-inch or four eighteen-inch torpedoes. This was also the first patrol boat design to install machine guns in power-boosted turrets.

Seeing the versatility of the British motor torpedo boat prototypes, the US Navy opened a call for proposals soliciting, among other sizes, a fifty-four-foot and a seventy-foot patrol torpedo boat. Specifications for the seventy-foot boats called for a vessel with a top speed of forty knots with a 275-mile radius, or a 550-mile radius at cruising speeds. The boat would have to reach its top speed while carrying a pair of twenty-one-inch torpedoes, four three-hundred-pound depth charges, and a pair of .50-caliber machine guns and ammunition. The competition was won by a design from the Sparkman and Stephens naval architecture firm that was to be built by Higgins Industries of New Orleans, Louisiana. The Sparkman and Stephens designs were designated *PT 5* and *PT 6*. Additional prototype PT boat contracts were subsequently let for PTs 1 and 2, both fifty-eight-foot designs awarded to the Fogal Boat Yard Inc./Miami Shipbuilding Co., Miami, Florida; PTs 3 and 4, also fifty-eight-foot designs fitted with Packard V-12 engines, were awarded to the Fisher Boat Works, Detroit, Michigan; and PTs 7 and 8 were awarded to the Philadelphia Navy Yard, Pennsylvania, for two boats designed by the Navy's Bureau of Ships.

Simultaneous to awarding the contracts for PTs 1–8, the boat that would become *PT 9* was acquired from the British Power Boat Co. by a representative from the Electric Launch Company (Elco), Groton, Connecticut. (Production Elco PT boats would be built at the company's Bayonne, New Jersey, shipyard.) This boat was shipped back to the United States as deck cargo on the freighter SS *President Roosevelt*, arriving in the United States on September 5, 1939.

After a pair of open-ocean "sail-offs" in July and August 1941, known as the "Plywood Derby," between PTs 1–9, the US Navy awarded a contract for a squadron of motor torpedo boats to Elco on December 7, 1939, calling for the delivery of *PT 9* plus eleven newly built PT boats. Higgins Industries finished *PT 5* and *PT 6* for

Left: When Motor Torpedo Boat Squadron 8 was decommissioned in October 1945, the boats were stripped of their armament and other useable parts and pulled up on the beach at Leyte. Identifiable in this photo are eleven boats from MTB-8, including *PT 114* (sixth from right), *PT 149* (third from right), and *PT 143* (far right), all of which are long-serving eighty-foot Elco boats placed into service in October 1942. Squadron commander Lt. William C. Godfrey had the unenviable task of ordering his boats destroyed, and that's exactly what happened. Considered war weary and uneconomical to ship back to the United States, the boats were splashed with gasoline and burned on October 28, 1945. *National Archives via Jerry Gilmartin*

the US Navy. Unhappy with their boats' performance, Higgins Industries used its own money to redesign *PT 5* to better meet the Navy's new specifications, and the company subsequently sold *PT 6* to the Finnish navy.

Moving forward, the US Navy settled on a new series of specifications for future PT boats: the boats were to be no shorter than seventy-five feet and no longer than eighty-two; power was to be provided by three Packard marine engines capable of muffling its exhaust noise for combat operations; have a cruising range of five hundred miles; and the PTs were to be capable of running at forty knots for one hour.

Just prior to America's entry into World War II, Elco, Higgins, and the Huckins Yacht Co. of Jacksonville, Florida, were awarded contracts for PT boats conforming to these new specifications. The boats built by Elco entered service in June 1942, Huckins Yacht Co. delivered its boats in July and August, and Higgins Industries made its first deliveries in September 1942.

It should be noted that shipyards in the United States license-built 140 of the Vosper seventy-foot design that were lend-leased to Britain, Canada, Norway, and Russia during the war. Although they were built in America, the US Navy operated none of the Vosper design boats.

The Huckins Yacht Co. built only two squadrons of PT boats (eighteen boats) before it left PT boat building to Elco and Higgins Industries. Huckins designed a balanced outboard rudder and licensed its Quadraconic hull design (sharp V at the bow) and laminated keel (which provides increased strength to the hull) to Elco and Higgins. With Huckins having limited production capacity, that left only two manufacturers building PT boats—a seventy-eight-foot hull boat built by Higgins Industries and an eighty-foot boat built by Elco.

HIGGINS BOAT WALK-AROUND

Industrialist Andrew Jackson Higgins had built a lumber importing business in the early 1920s. To facilitate his business, Higgins acquired a large fleet of ships to transport his lumber. Out of necessity, in 1930 he opened a shipyard called Higgins Industries to maintain his fleet and to construct new ships for the lumber, oil, and gas businesses in the area, and to build new ships for a number of Coast Guard contracts.

In addition to winning the contract to build seventy-eight-foot PT boats for the Navy after America's entry into World War II, he also built a huge number of landing craft—both LCVPs (Landing Craft, Vehicle, Personnel) and LCMs (Landing Craft, Mechanized), the latter of which were capable of bringing a tank ashore. In total, Higgins Industries built 199 motor torpedo boats in three classes (*PT 71*-, *PT 235*-, and *PT 625*-class boats).

Seeing a PT boat in the water for the first time, one is struck by the deep V of the hull and how the bow sits very tall in the water. Stepping aboard at the bullnose, the bronze semi-circle that guides ropes and the anchor chain, visitors are greeted by a large variety of armament. This boat is built for one purpose, and every inch of the main deck is used for offensive armament. At the bow is a thirty-seven-millimeter M9 cannon built by Oldsmobile, the same cannon found in the US Army Air Forces' Bell P-39 Airacobra and P-63 Kingcobra single-seat fighters. The M9 has a unique ammunition feed magazine, one that comes over the top of the gun as used in the fighter aircraft.

On the port side of the forward deck is a twenty-millimeter Oerlikon cannon that can be used for land and sea targets or elevated to fire at low-flying aircraft. Situated on the centerline of the main deck aft of the M9 cannon is the twenty-millimeter and thirty-seven-millimeter ready ammunition locker and, to its rear, mounted forward on the port side of the chart house, is the spare twenty-millimeter barrel box.

Attached to the starboard side of the face of the chart house is the life raft and survival equipment. On each side of the boat, forward of the chart house and behind the twenty-millimeter and thirty-seven-millimeter cannon, are two eight-barrel Mk 50 rocket launchers that fire five-inch-diameter, spin-stabilized rockets that have an effective range of eleven thousand yards. The Mk 50 rocket launchers pivot, extending over the side when in firing position, and sit over the deck when stowed.

On the backside of the chart house is the open bridge with the helm and engine controls located on the port side and the BK transponder antenna (Identification Friend or Foe, or IFF, system—BN interrogates other systems and BK sends out the PT boat's IFF code) and searchlight on the starboard side. Located between the helm and searchlights is the hatchway leading below deck. On each side aft of the bridge are two, twin Browning M2 .50-caliber air-cooled machine guns in Bell Mk 9 gun cradles in Elco Mk 17 turrets, and on the centerline is the Raytheon Corp. SO3 surface search radar (ten-centimeter wavelength; twenty-nautical-mile range) radome with BN antenna mast. Directly behind the mast is the engine room hatch, which has an opening for personnel or, when fully opened, is large enough to lift or lower an engine through. Behind the engine room hatch is another twenty-millimeter Oerlikon cannon, the forty-millimeter ready ammunition locker, and the forty-millimeter Bofors cannon, and the boat's smoke generator sits above the stern.

Along the rail on each side of the boat are two Mk 13 aircraft torpedoes, each fitted with a six-hundred-pound warhead in Mk 1 launcher racks (roll-off racks, four in total). When a torpedo's steam engine is brought up to speed, wire cables are released and the thirteen-foot, six-inch-long, twenty-two-inch diameter underwater missile rolls off the sides of the boat and into the water. Behind the torpedo racks is a Mk VI three-hundred-pound depth charge, one on each side.

Entering the forward below decks area through the bridge hatch, a PT boat crewman would drop down two stairs to the chart house. To the right is the chart table and on the left side is the boat's radar screen. Radio communication equipment is located on the forward bulkhead and shelf (level with the chart table and radar receiver). Below the radio shelf, a companion way leads down stairs to the crew's quarters. The gallery and forward head are located in the crew's quarters, with four berths on each side of the compartment when facing forward. At the bow of the boat at this level is the Bosun's Locker, and in the middle of the deck is a multi-use table and a ladder that leads to a "Booby Hatch" overhead that exits on deck behind the thirty-seven-millimeter cannon.

Turning around and walking aft through two watertight doors (one door from the forward crew quarters and the other into the officers' head) is the wardroom. The captain's berth is on the boat's port side, and the second officer's berth is situated on the starboard side. There is a small writing desk and chair next to the captain's berth, and through the aft-facing bulkhead is the officers' head. On each side of the lavatory is an eight-hundred-gallon fuel tank.

The engine room cannot be accessed through the crew's quarters and wardroom; one must descend a ladder from the main deck into the compartment. At the bottom of the engine room ladder are three 1,850-horsepower, twelve-cylinder (60° V configuration) Packard-Marine 5M-2500 engines with one each on the port and starboard side and the third situated to the rear on the boat's centerline. All three engines drive their own propellers, and the starboard and port propellers are positioned ahead of the rudders for improved handling.

Behind the engine room is the aft crew's quarters with six berths and the Lazarette where the steering gear and main rudder gearbox are located. If rudder control was knocked out, a seaman could disconnect the steering linkage, install an eight-foot-long "emergency tiller bar" onto the steering linkage, and manually control the rudders from the main deck. Additional lines, fenders, and other equipment are stored here as well.

THE LIFE AND TIMES OF *PT 658*

Today, there are fourteen Patrol Torpedo boats left out of 531 that served during World War II (see Surviving PT Boats, page 206). One of the surviving boats is *PT 658*, built at Higgins Industries in 1945. In April of that year, *PT 658* was assigned on paper to MTB Squadron 45 and was slated to be lend-leased to the Soviet Union. Higgins Industries completed the boat on July 30, 1945, nearly three months after Germany surrendered, and by this time the United States had stopped sending lend-leased military equipment to its allies.

After extensive tests on Lake Ponchartrain, the Board of Inspection and Survey sent a letter accepting *PT 658* for delivery to the Navy on July 30, 1945. *PT 658* and three other boats (*PT 657, 659,* and *660*) at Higgins Industries were then loaded onto the deck of *LST-375* and transported to their first assignment at Bremerton, Washington, arriving on September 25, 1945. A few weeks later, *PT 658* was subsequently transferred to the Pacific Missile Test Range at Point Mugu, California (sixty-five miles northwest of Los Angeles), and stationed at the facility's harbor

Above and left: Two views of *PT 631*, a *PT 625*-class motor torpedo boat and sister ship to *PT 658,* idling near the Higgins Industries factory in New Orleans, Louisiana. *Naval Historical Center*

PT 657, PT 658, PT 659, and PT 660 were loaded onto the deck of LST-375 and delivered to the naval base at Bremerton, Washington, in September 1945. The boats were only at Bremerton for a few weeks before they were shipped to Port Hueneme in Southern California. *National Archives via Jerry Gilmartin*

SPECIFICATIONS: *PT 658*

Length	78 feet 6 inches
Beam	20 feet 1 inch
Draft	5 feet 3 inches
Displacement	48 tons
Powerplant	3 2,490-cubic-inch 5M-2500 Packard-Marine V-12 engines producing 1,850 hp each at 2,500 rpm
Crew	2 officers, 14 enlisted
Armament	1 M9 37-millimeter Oldsmobile cannon 2 M4 20-millimeter Oerlikon cannon 4 .50-caliber machine guns in twin gun mounts 4 Mk 13 torpedoes (600-pound warhead) with a range 6,300 yards 2 Mk VI depth charges (300 pounds of TNT) 2 Mk 50 rocket launchers firing Mk 7 or Mk 10 5-inch-diameter spin stabilized rockets with a range of 11,000 yards
Launched	July 30, 1945
Builder	Higgins Industries
Preserved by	Save the PT Boat Inc.
Website	www.SaveThePTBoatInc.com

On November 7, 1949, *RCT-13*, *RCT-14*, and *RCT-15* were photographed at their home base of Port Hueneme, part of the Pacific Missile Test Range at Point Mugu Naval Air Station. *RCT-14* (closest to the camera) has a target reflector sitting on both the bow and the aft deck, while *RCT-15* (center) and *RCT-13* are in the patrol configuration. *RCT-13* was delivered to the US Navy as *PT 658* on July 30, 1945. When sold surplus in June 1958, the boat looked very much like it does in this photograph. *Save the PT Boat Inc. collection*

known as Port Hueneme. On August 27, 1948, *PT 658* was reclassified as a "small boat" and its identity changed to C105343. As a small boat, she would patrol the waters of the missile test range to ensure that the area was clear of civilian boaters, and she would act as a target for the forthcoming generation of missiles and radars then under development. In December 1948, the boat was reclassified as *RCT-13* for patrol use and to take crews to and from the early warning radar station on nearby Santa Rosa Island.

After a dozen years of service at Port Hueneme, *PT 658/RCT-13* was sold as surplus equipment on July 30, 1958. The boat was purchased by Earl C. Brown, proprietor of Brown Marine Salvage in Oakland, California, for $1,201 in "as is" condition without engines. Brown renamed the former motor torpedo boat *Porpoise* and worked to convert it into a pleasure craft by installing a pair of 485-horsepower Detroit Diesel 6-71 marine engines (six cylinders with each displacing seventy-one cubic inches). Although the 6-71s burned less fuel, the difference in horsepower and the fact that there were only two of the diesels versus three of the 2,500-horsepower Packard engines made the boat slow, and it wallowed in the water due to its displaced center of gravity. Apparently, Brown took the boat out on the San Francisco Bay a couple of times and everyone on board became seasick breathing diesel fumes while the boat Dutch-rolled across the water. After one final trip, Brown tied the boat to a dock in the Alameda Estuary, and it reportedly never moved again.

A BOAT ADRIFT IN SAN FRANCISCO BAY

In 1991, San Francisco Bay Area resident, and former *PT 231* skipper, Ed Jepson recognized the hulk as an old Higgins PT boat. Jepson contacted Oregon Military Museum curator Terry Aitken and PT boat veteran Harry Weidmaier, who immediately began correspondence with Brown Marine Salvage Corp. to obtain the boat.

In 1992, Earl Brown passed away and his son, Orlando, inherited the boat. The former PT boat had sat, tied up in the Alameda Estuary, untouched for years and suffering from exposure to ultraviolet rays from the sun. And although the sun gives life to many things, it had rotted most of the exterior wood and deck planking. The boat was a mess.

After a couple of tries at negotiations with Orlando Brown, the PT boat veterans were able to come to an agreement. On the condition that the veterans would restore

This photo is from the 1992 trip to determine *PT 658*'s condition and plan its future. Boat-slip rental fees being what they are, the veterans from Save the PT Boat Inc. soon moved the diamond in the rough to the Clipper Cove Marina at the Treasure Island Naval Base in the heart of the San Francisco Bay. *Save the PT Boat Inc. collection*

the weather-beaten boat to its original condition, Orlando Brown sold *PT 658* to the veterans who had, by this time, formed the non-profit corporation "Save the PT Boat Inc." of Portland, Oregon. The sum of $1 changed hands, and the veterans now had to find a way to move their prize possession.

PT 658 was moved from the Alameda Estuary to a dock at the Clipper Cove Marina (so named for the Pan Am Clippers that flew from this location to the Orient in the mid-1930s) on Treasure Island, located halfway between San Francisco and Oakland, and the site of the former Treasure Island naval base. Veterans and volunteers would travel from the Portland area to Treasure Island to work on the boat as they prepared it for its eventual move north.

During a particularly bad storm during December 1992, *PT 658* broke away from its dock and drifted into shallow water. During the storm, the starboard bottom of the boat was holed and the boat began to sink. The only thing keeping *PT 658* afloat was the more than three thousand gallons of rotten diesel fuel in her tanks, and this was only holding the boat's chart house a few inches above the water. At the time, Treasure Island was still an active Navy base, and tugs and other military watercraft would travel between the island and the Naval Air Station at Alameda, less than three miles away by boat. As a Navy tug was passing by Clipper Cove, the sharp-eyed chief at the helm saw the PT boat with its decks awash. He maneuvered his boat behind *PT 658* and pushed it up onto Clipper Cove's only sandy beach at high tide. The tug's skipper then phoned the veterans, who raced down from Portland to take stock of their boat.

During a storm in December 1992, *PT 658* broke from her moorings when the bullnose and a deck cleat were torn off, casting the boat to the currents. Her stern hit the breakwater and caved in the starboard rudder, making an eight-by-ten-inch hole in the boat's bottom. Flooded with water, the diesel fuel in her tanks was all that kept *PT 658* from sinking to the bottom. *Save the PT Boat Inc. collection*

A Navy chief piloting a tug between the Navy bases at Treasure Island and Alameda spotted the nearly sunken PT boat and pushed her up on the beach. He phoned the veterans group, which put together a salvage party to rescue the historic boat. When the tide receded, it gave the veterans the opportunity to patch the hole and pump out the boat. *Save the PT Boat Inc. collection*

Upon arrival, they found the PT boat's lines still tied to the dock. What was unusual was at the other end of the lines: the boat's bullnose and a deck cleat were still tied to the lines, but they had been ripped from the rotted deck in the storm, setting the PT adrift. It was a sad sight to see.

When the tide went out, PT boat veterans were able to see the damage. A rock had pushed the starboard rudder up and into the boat, causing an eight-inch by ten-inch hole in the bottom of the boat. The boat was patched, pumped out, and prepared for the forthcoming journey to Save the PT Boat's home in Portland.

BACK FROM THE BRINK

Sometimes it takes the sum of two parts to make one whole. That's the case with *PT 658*.

In the 1980s, the US Navy loaned *PT 658*'s sister ship, *PT 659*, to the Oregon Military Museum at Camp Withycombe in Clackamas County, south of Portland. *PT 659* was delivered on an original, World War II-vintage PT boat cradle, and it had its original engines still installed. The Navy operated *PT 659* until 1970, and through the years the boat had been highly modified and was not as original as *PT 658*. The Oregon Military Museum wanted to transfer *PT 659* to the veterans' group, and they were quite interested. Presented with a choice between the two, Save the PT Boat Inc. chose to restore *PT 658* and pull parts from *PT 659*. After they secured the parts they needed, *PT 659* was then transferred to the PT Boat Council of Vancouver, Washington, a group that planned to restore the boat as a land-based static display.

From *PT 659*, the veterans' group swapped cradles, keeping the original one for *PT 658* and setting *PT 659* on a custom-built stand. *PT 659*'s 1945-vintage Packard 5M-2500 gas engines were removed and set aside for refurbishing for *PT 658*. Although *PT 658* was originally delivered with 4M-2500 engines, the new 5M-2500 engines were of the same vintage, and the only difference was that they were more powerful, with an intercooler and a larger supercharger. Had the war continued, *PT 658* would most likely have received the up-rated engines, so the veterans' group was more than pleased to acquire them.

With the cradle from *PT 659* in their possession, it was quickly sent south, and *PT 658* was lifted from the waters of the San Francisco Bay and set on dry land. While it was high and dry, a number of temporary repairs were made to the boat to improve the working conditions for volunteers. A new plywood deck covering was added for safety reasons, and the boat was cleaned inside and out. By 1993, it was ready to be moved to Portland. But how?

"Harry Weidmaier [captain, USN, ret.], who was Save the PT Boat's second president, was a retired Navy captain, and he was the one who was able to arrange for the Washington Army National Guard to move the boat for us," said Jerry Gilmartin of Save the PT Boat. "The Washington National Guard had a ship stopping at the Alameda naval base that was capable of moving our boat, and Weidmaier negotiated to get *PT 658* moved as part of a training exercise."

The National Guard's ship was the USAV *General Brehon B. Somervell* (LSV-3), a logistics support vessel assigned to the 805th Transportation Detachment (US Army Reserves), home-ported at Tacoma, Washington. *Somervell* is much like an LST (Landing Ship Tank) of World War II fame with the exception that it has an open well-deck capable of carrying fifteen M1 Abrams main battle tanks (each

As the restoration got started, Save The PT Boat Inc. acquired the Packard Marine engines it needed to restore *PT 658* from *PT 659*. In addition to yielding parts for the restoration of *PT 658*, *PT 659* gave up substantial sections of its hull to restore *PT 305*, a combat veteran PT boat under restoration at the National World War II Museum in New Orleans, Louisiana. *Save the PT Boat Inc. collection*

twenty-six feet long by twelve feet wide). While most of *Somervell*'s cargo is typically tanks, trucks, or containers, the open-top well deck enables large, outsized loads to be craned aboard from overhead. *PT 658*'s seventy-eight-foot length and twenty-foot width took up a space two tanks wide and four tanks long for a total footprint equal to eight M1 tanks. Thus, with the patrol boat on board, *Somervell* still had plenty of room left to carry other cargo required for its mission.

For *Somervell* and its crew, moving the PT boat was considered a training mission, but one that would not take the ship from its intended route. Because of this, *PT 658* was first moved to *Somervell*'s homeport at Tacoma, where the patrol boat sat for six months. When a mission to deliver cargo to the Portland area came up, *PT 658* was loaded on *Somervell* and delivered to its new home in June 1994. At Portland, the Army Corps of Engineers used a fifty-ton stiff-leg derrick with a seventy-two-foot boom at Mooring Dock A on the Willamette River to lift *PT 658* from *Somervell* onto a Navy-surplus barge, which would become the torpedo boat's home while it underwent restoration during the next decade.

"The barge and boat were moored to a dock at the US Navy and Marine Corps Reserve Readiness Center on Swan Island," said Gilmartin. "The first thing the group did was replace the deck. Then they had to rebuild the runners and the steering gear, which was damaged when the boat got loose in Clipper Cove. Then they completely stripped the engine room and removed all the wiring in the boat because it was rotten."

While the initial steps were being taken to preserve the boat and assess what needed to be done to repair her hull, the three Packard 5M-2500 engines were taken to Camp Withycombe to be overhauled. "Terry Aitken was very generous in helping to facilitate the boat's restoration by providing space for our group to work on components, like the engines, indoors and out of the weather," said Gilmartin. "At the time, Jim Brunette was in charge of the restoration. He kept a daily log of everything that was done, and today we're able to draw on his records, which make our maintenance tasks so much easier. His records are amazing."

Having the boat in the cradle enabled the restoration crew to replace all of the gusset plates with aluminum units. In order to replace the gusset plates, the bottom two strakes of hull planking on the outside of the boat had to be removed, and this gave the restorers the opportunity to replace the planking with new material. They were also able to remove and replace any rotten wood on the hull. The bottom was in good shape; most of the rotten wood was on the sides of the hull.

When the deck came off the boat, it was obvious that the chart house would need to be rebuilt. "Don Brandt removed the chart house, carefully taking it apart piece by piece," said Gilmartin. "He took the entire thing to his house and replicated each piece exactly using new wood. He put the new chart house together and everyone was extremely impressed. Whether you realize it or not, your eyes are drawn to the chart house as the main feature of the boat. It's such an iconic piece when you consider that nearly all photographs of John F. Kennedy on board *PT 109* show him standing at the helm or his crew assembled around the front of the chart house. And most PT boat movies and TV shows spend a lot of time interacting with the crew on the open bridge. The rebuild of the chart house had to be perfect, and it was."

Once the hull was sound, it was time to reinstall the boat's running gear. The Marine Corps crane on the pier alongside the boat was used to lower the Packard

5M-2500s through the deck hatch and into the engine room. Once the engines were installed, the propeller shafts were fitted. Each shaft is two and a quarter inches in diameter and is made of Monel, an alloy with high nickel content, which is extremely strong and resistant to corrosion. Then the rudders were reinstalled and connected.

The boat was being painted in 2003 when there was a huge setback, resulting from a small on-board fire in the officers' wardroom caused by an electrical short in an auxiliary lighting system. There was a US Navy ship moored nearby, and one of its sailors on duty noticed smoke coming from the PT boat. The fire department was called and the fire put out, but not before the entire wardroom was gutted and one of the fuel tanks was holed during the firefighting process. It was a setback, and at the time it looked like it might be a restoration-ending problem. A shipwright was hired, and working with the volunteer crew they replaced a couple of longitudinal beams in the deck of the officers' wardroom, the entire ammunition locker (situated between the crew's compartment and the wardroom), and the plywood deck above the officers' stateroom. Bob Alton volunteered to climb into the fuel tank to seal it up. Repairing the damage took six months and consumed money that could have been better spent elsewhere, had there not been a fire.

Now seaworthy with running engines and many of the armament pieces installed, *PT 658* was ready to go back into the water. Everyone had hoped for June 6, 2004, the sixtieth anniversary of the D-Day invasion, but things took a day or so longer. The restored PT boat made its first public appearance at the Antique and Classic Wooden Boat Show on the Willamette River during the last weekend of June 2004. The patrol boat received a tremendous welcome from the general public and from boating enthusiasts, who understood the amount of effort that was put into the restoration.

PT 658 IN ACTION

PT 658 has attracted a lot of attention by sailing the Willamette River, and each year the authenticity of the restoration improves. The most visible change came in 2013 when the boat was painted in its factory delivery colors, known as World War II Measure 31, Design 20L camouflage, which consists of three shades of green and black in various patterns.

During the sailing months, from June to the end of September, the boat can be seen at various nautical events in the Portland area. To get *PT 658* from dock to destination takes a highly skilled crew and one that is cross-trained to operate all the boat's different systems. For those working the engine room, the heat and noise put out by the three V-12 marine engines requires that a sailor can work for only sixty minutes before he or she has to be relieved for a rest period.

Descending the engine room ladder puts sailors into a stark white room, dominated by three dark gray, twelve-cylinder Packard Marine 5M-2500 engines. The port engine is called "Bob's" for Bob Alton, the starboard engine is "Tom's" for Tom Cates, and the center engine is "Bubba's" for Daniel "Bubba" Conway, each a long-serving *PT 658* volunteer who is responsible for maintaining the engines. And all three engines are mighty thirsty. At today's prices it costs $10,400 to fill *PT 658*'s tanks, and all three Packard V-12 engines burn five hundred gallons per hour at top speed (approximately forty knots). To keep fuel costs down and to save wear and tear on the boat, Save the PT Boat limits its top speed to a maximum of twenty-three

The helm of a *PT 625*-class motor-torpedo boat as delivered from the Higgins Industries factory and *PT 658*'s restored helm today. Any of the original components that could not be located were replicated by Save the PT Boat's volunteer craftspeople. *Then: Naval Historical Center; restored: Nicholas A. Veronico*

knots (26.5 miles per hour). In addition, the boat puts out a tremendous wake at higher speeds, and cruising at a slower pace keeps those along the banks of the Willamette River much happier. Being a good neighbor is part of the safe operation of a PT boat as well.

Tom Cates, a US Marine Corps veteran who retired from the communications industry and is the namesake for the boat's starboard engine, said:

> I've been at it for a long time. Our guys are just motor vehicle backyard mechanics, and we know how to read books and have mechanical sense. This is exactly what

they'd look for in World War II. They looked for farm boys that could fix a John Deere with bailing wire and a pair of pliers in the middle of the night because that's exactly what they had to do on patrol. You have to figure out the engines. We didn't have any books to start with other than what the World War II motor mechanics put together from memory. Then we had the opportunity to work alongside many PT boat veterans who were recalling tasks they performed more than forty years ago. Some things we got right, and others we didn't, but the boat runs well and we've got a good crew.

They are very simple boats that if you try to make it too complicated . . . just [give] you more things to fail. PT boats were designed to keep everything just as simple as possible and there's always redundancy; for example, if one engine craps out, you've got another engine to back you up. The boat can operate on just one engine, but it's more maneuverable with all three working, especially the wing engines. The wing engines drive propellers that are directly in front of the rudders, which gives you steering capability.

As far as getting down in the engine room on cold, rainy days, it's a pretty good place to work. You wear ear protection because at idle, the wing engines put out about 105 decibels, and if you've got any rpms on all three engines, it's well over 110. In the heat it usually averages about ninety degrees in the engine room,

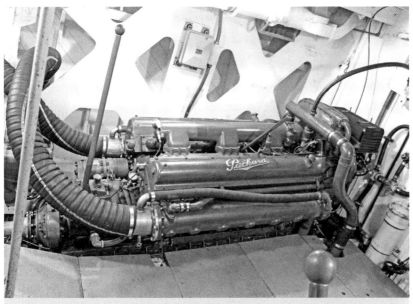

The portside twelve-cylinder Packard Marine 5M-2500 engine, known to the crew as "Bob's" engine after Bob Alton, Save the PT Boat's president and the volunteer responsible for the maintenance and upkeep on this engine. The cooling-jacket tubes are six-inch-diameter stainless-steel pipes, which exhaust the engines underwater on both the port and starboard sides and are covered with a rubber hose. Cooling water passes between the stainless-steel pipe and the rubber exterior hose to keep the external temperature down. Working together, these were custom made by the Albina Pipe Bending Company in Tualatin, Oregon, and Premier Rubber and Supply in Portland. *Nicholas A. Veronico*

The fully restored *PT 658* idles on the Willamette River near the city of Portland, Oregon. All of the armament has been replaced on the deck, and the starboard forward Mk XIII torpedo is a cutaway, enabling docents to explain the inner workings of the underwater missiles to guests. The boat wears its original World War II Measure 31, Design 20L camouflage, which consists of three shades of green and black in various patterns. Each year, she sails between June and the end of September. *Wally Boerger*

sometimes higher in the summertime. After a while, the heat and the noise make you kind of rummy. You are only down there about an hour or so, then somebody else spells you. During patrols in World War II, the gunners would relieve the engine mechanic.

Our primary duties are to monitor the engines and shift them when ordered to do so by the skipper. Each engine is autonomous, so it has its own set of gauges, its own heat exchangers, and its own gearbox. At the top of this gauge board is an arrow board [similar to the pointer on an engine order telegraph] that is cabled to the helm. The skipper will give you a signal to shift the port or starboard engine forward or aft, and that arrow tells you which direction to shift the engines. The skipper gives you a signal before he's going to do something using an air horn we have in the engine room.

You are watching the arrow board and when the arrow moves you'll shift the engine selected by the skipper to the direction indicated. The skipper waits about three seconds to give you time to make the change, and then he has access to

the throttle at the helm. There are provisions to answer that the change has been made, but usually when you work together for a while, the skippers can feel what you are doing and it's just practice, practice, practice. That's the way the boats were run during the war, and we're doing the same thing today.

FUTURE PLANS

When they formed the group, preserving a PT boat was a distant dream. Today, Save the PT Boat has succeeded beyond many of the veterans' and other volunteers' wildest dreams. The organization is working toward securing its future by raising $3.5 million for the construction of the *PT 658* Education and Heritage Center. The new building, located on the Willamette River along the Portland waterfront, would enable thousands of people to see and tour the boat each year, allow for maintenance in a covered dock, and provide room for an archive/library within the heritage center. The center will be fantastic tribute to those who built, sailed, and restored the "Devil Boats."

SURVIVING PT BOATS

Hull No.	Builder	Owner or Operator	Status*	Website	Notes
PT 3	Fisher	Private owner, New Jersey	S	n/a	In storage at Flanigan Brothers Boatyard, Fairton, New Jersey; reported for sale in 2012
PT 8	Fisher	Private owner, Franklin, Louisiana	S	n/a	Only all-aluminum construction PT boat built; reported for sale
PT 48	Elco	Fleet Obsolete, Kingston, New York	S	www.fleetobsolete.org	Extensive combat history in the Pacific Theater; currently for sale
PT 305	Higgins	National World War II Museum, New Orleans, Louisiana	R	www.nationalww2museum.org	Under restoration, slated for completion in May 2015.
PT 309	Higgins	National Museum of the Pacific War, Fredericksburg, Texas	PV	www.pacificwarmuseum.org	On public display; externally restored
PT 459	Higgins	Fleet Obsolete, Kingston, New York	S	www.fleetobsolete.org	D-Day combat history; currently for sale
PT 486	Elco	Fleet Obsolete, Kingston, New York	R	www.fleetobsolete.org	Reportedly under restoration
PT 615	Elco	Fleet Obsolete, Kingston, New York	R	www.fleetobsolete.org	
PT 617	Elco	PT Boats Inc. (displayed at Battleship Cove Naval Museum in Fall River, Massachusetts)	PV	www.ptboats.org	Now indoors at Battleship Cove (www.battleshipcove.org)
PT 657	Higgins	Malihini Sportfishing, San Diego, California	OW	www.malihinisportfishing.com	Converted and now operates fishing charters
PT 658	Higgins	Save the PT Boat, Inc., Portland, Ore.	OW	www.savetheptboatinc.com	Restored and operates each sailing season
PT 724	Vosper	Liberty Aviation Museum, Port Clinton, Ohio	S	www.libertyaviationmuseum.org	Converted to yacht configuration; awaiting restoration
PT 728	Vosper	Liberty Aviation Museum, Port Clinton, Ohio	R	www.libertyaviationmuseum.org	Restoration underway
PT 796	Higgins	PT Boats Inc. (displayed at Battleship Cove Naval Museum in Fall River, Massachusetts)	PV	www.ptboats.org	Now indoors at Battleship Cove (www.battleshipcove.org)

PT BOAT UNCONFIRMED SURVIVORS, RUMORS, AND WRECKS

Hull No.	Builder	Owner or Operator	Status*	Website	Notes
PT	Higgins	Ushuaia, Argentina, *Ara Towwora*		n/a	Reported on shore
PT	Higgins	Rio de la Plata, Argentina		n/a	Reported tour boat
unknown	Vosper	United Kingdom		n/a	Status unknown
unknown	Vosper	United Kingdom		n/a	Status unknown

* S = in Storage; R = under Restoration; OW= On the Water and sailing; PV = displayed on Public View

HIDDEN IN
THE NESTS

MANY OF TODAY'S SURVIVING WORLD WAR II–ERA ships escaped the ravages of war and then avoided the scrapper's torch by being hidden in the mothball fleets. They sat in the mothball fleet, waiting for the right time and the right people to restore them to their former glory.

The Merchant Ship Sales Act of 1946 sold off a large number of ships built for the war, and at its peak held more than 2,277 vessels in case they were needed to meet the demands of a future emergency. Those ships held for future use were managed under the Maritime Commission's National Defense Reserve Fleet. In 1950, the Maritime Commission was reorganized into the Maritime Administration (MARAD), which today is part of the Department of Transportation.

The initial reserve fleets were held in eight locations: Mobile, Alabama; Benicia, California; Stony Point, New York; Wilmington, North Carolina; Astoria, Oregon; Beaumont, Texas; Fort Eustis, Virginia; and Olympia, Washington. Through the years, the National Defense Reserve Fleet has held a variety of ship types, everything from Liberty and Victory ships, to icebreakers and oil tankers, to the once super-secret Hughes *Glomar Explorer* and the HMB-1 barge used to recover a Russian submarine, as well as the proof-of-concept *Sea Shadow* stealth ship.

By the end of October 2014, MARAD's reserve fleet only held 109 vessels. Of those 109, only three were ships built during World War II—a pair of buoy tenders (WLB-307 and -395) and *Sturgis* (MH-1a), a nuclear power station built on the hull of the World War II Liberty ship *Charles H. Cugle* (built at J. A. Jones Construction Co., Panama City, Florida, August 13, 1945). The reserve fleets of today hold roll-on/roll-off cargo ships, crane ships, and break bulk carriers. Gone are the vast fleets of World War II–vintage ships.

The most recent example of a World War II ship removed from the mothball fleet to become a floating museum is *Iowa* (BB-61), now cared for by the Pacific Battleship Center and displayed in Los Angeles Harbor.

Iowa was laid down on June 27, 1940, at the New York Naval Shipyard, Brooklyn, New York, and launched twenty-six months later on August 27, 1942. *Iowa's* sponsor was Mrs. Henry A. Wallace, wife of the then-serving, thirty-third vice president of the United States. The ship was commissioned in February 1943 and began its first war patrol on August 27, headed for the North Atlantic. The battleship carried President Franklin D. Roosevelt to the Tehran Conference in November and December of that year. In preparation for this mission, *Iowa* gained a square bathtub for Roosevelt's use in the Captain's In-Port Cabin on the 01 Level (first deck above the main deck).

In 1944, *Iowa* was transferred to the Pacific theater, and sailing with the battleship *New Jersey* (BB-62) became part of Task Force 58. Ships of the task force shelled enemy-held islands across the Pacific, from Kwajalein and Eniwetok Atolls to Ponape, Saipan, and Tinian. The next year, *Iowa* rained sixteen-inch shells down on the Japanese home islands from Okinawa to Honshu to Hokkaido. During the Japanese surrender on September 2, 1945, *Iowa* served as the flagship for Adm. William "Bull" Halsey.

Battleship *Iowa* was mothballed in 1948, and in July 1951 was recalled for duty in the Korean War. Her sixteen-inch guns again wreaked havoc on shore targets; in this conflict they focused on industrial plants and railroad infrastructure—tracks, tunnels, and bridges. After the Korean War, *Iowa* served the US Navy until she was inactivated in February 1958, and stored in the Atlantic Reserve Fleet at Philadelphia, Pennsylvania.

Iowa sat out the Vietnam War and was called up in 1982 to be modernized to serve as a Cold War deterrent. Fitted with launchers for BGM-109 Tomahawk cruise missiles and AGM-84 Harpoon anti-ship missiles, as well as the Phalanx antiaircraft and antimissile defense system, BB-61 sailed in support of the US Navy's commitment to the North Atlantic Treaty Organization (NATO).

In April 1989, during a gunnery exercise, there was an explosion in turret number two, which took the lives for forty-seven sailors. After a shameful and misguided

On October 28, 2011, *Iowa* was towed from Benicia through the Carquinez Straits and San Pablo Bay and into the Richmond Harbor, where she would undergo the beginning of her restoration. Notice that the radar mast had not been installed at this point. *Nicholas A. Veronico*

Right: An aerial view shows *Iowa* in the Suisun Bay mothball fleet in summer 2007. Boaters traveling from San Francisco to Stockton and Sacramento were able to get a close-up view of the historic ship as they cruised through Suisun Bay. *Nicholas A. Veronico*

Iowa passes under the Golden Gate Bridge for the final time on May 26, 2012, en route to Los Angeles. The battleship was escorted from Richmond Harbor to the Golden Gate by a huge flotilla of pleasure craft and a fireboat offering a two-hundred-foot-tall salute in the form of water jets. *Nicholas A. Veronico*

attempt to blame one of the deceased sailors for the explosion, an independent investigation by the Sandia National Laboratory, Alamogordo, New Mexico, determined that the powder bags were over-rammed into the breech at high speed, igniting the shell's propellant, which resulted in an open-breech explosion.

The battleship was then decommissioned in October 1990. The subsequent fall of the Soviet Union in December 1991, and the Navy's ability to deploy cruise missiles from smaller, more economical-to-operate ships, ended the need for battle-wagons. *Iowa* was tied up at the Naval and Education Training Center in Newport, Rhode Island, from 1998 to 2001, when it was towed to the mothball fleet in Suisun Bay at Benicia, California, in hopes of one day establishing a naval museum in San Francisco. That idea was a political hot potato, and it was quickly shelved. In March 2006 *Iowa* was struck from the Naval Vessel Register, and the move to find the battleship a new home began.

Repainted and ready to welcome guests, *Iowa* rests in the Richmond Harbor before being towed to Los Angeles for display. It took three thousand gallons of paint to make the battleship presentable. *Roger Cain*

Two groups, one in Stockton and the other in Vallejo, California, attempted to make a home for *Iowa*, but both groups were unable to make that dream a reality. In May 2010, *Iowa* was offered to viable nonprofit groups interested in turning the battleship into a museum and memorial. On the last day for submissions, the Pacific Battleship Center turned in its proposal. Nearly one year later, on September 6, 2011, the secretary of the Navy awarded *Iowa*'s custody to the Pacific Battleship Center. Wasting no time, on October 27, *Iowa* began the two-day move from the mothball fleet to a pier at nearby Benicia. The following day, when the tides were right, the battleship was moved through the Carquinez Straits into the San Francisco Bay, and to the harbor at Richmond, California, where the exterior of the ship was refurbished, making BB-61 display ready.

Bay Ship and Yacht Co., of Alameda, California, used more than three thousand gallons of paint to cover approximately 210,000 square feet of *Iowa*'s exterior, and on April 24, 2012, the mast radar platform was "re-stepped." This event called for a special ceremony with honored guests from the State of Iowa on hand to witness the re-stepping. The thirty-seven-thousand-pound, fifty-foot-tall radar mast was lifted and bolted into place, giving the ship an overall height from the waterline to the top of the mast of 174 feet.

On May 20, *Iowa* was slated to be towed out of San Francisco Bay to her new home in the Port of Los Angeles, but stormy weather along the route postponed the voyage. Six days later, on May 26, the tug *Crowley Warrior*, assisted by three other tugs, pulled the World War II–era battleship past saluting fireboats, hundreds of small craft, and out under the Golden Gate Bridge into the open water of the Pacific Ocean.

Arriving at its permanent home on June 9, 2012, in the Port of Los Angeles at Berth 87, a series of special celebrations began as *Iowa* was readied to become a museum. The ship was opened to the general public on July 7, 2012. As volunteers devote time to refurbishing the ship's interior spaces, more and more areas are opened to the public. Today, the Battleship *Iowa* tour takes guests from the second

Now tied up in Los Angeles Harbor at Berth 87, facing the hills of San Pedro, *Iowa* is open for tours while a dedicated cadre of volunteers works to restore the battleship. This is certainly the most popular place on the ship to take a photo. The sight of six sixteen-inch/50-caliber gun barrels is extremely impressive and showcases *Iowa*'s firepower. The Vincent Thomas Bridge can be seen in the background. *Roger Cain*

The guns of the after turret (turret no. 3) have been elevated to allow for increased display space on the aft deck. Each gun is 66.6 feet long, which is fifty times greater than their sixteen-inch bore (thus the term 50-caliber). Of the gun's overall length, a 43-foot portion of the barrel is visible outside the turret. Each of the hatches leading below deck is covered as shown here. *Nicholas A. Veronico*

deck to the 05 Level, through such historic spaces as the Captain's In-Port Cabin, the Signal Bridge, the Flag Bridge, the crew's quarters, the Officers' Wardroom, and past the gun turrets, Tomahawk missile launchers, and Phalanx defense systems.

Although she'll never sail again under her own power, *Iowa* today serves as a monument to all those who served in the US Navy, and represents naval power from World War II, Korea, and the Cold War.

MERCHANT SHIP SURVIVORS

What if one could touch a living, breathing World War II armed merchant ship? What would it be like to watch the engine turn, see how the machinery works, or even sail on one? Through the efforts of thousands of volunteers, the opportunity

SPECIFICATIONS: USS *IOWA* (BB-61)	
Length	887 feet 3 inches
Beam	108 feet 2 inches
Draft	37 feet 9 inches
Displacement	45,000 tons
Powerplant	General Electric geared steam turbines capable of 212,000 shaft hp driving 4 propellers
Top speed	32.5 knots
Crew	2,700 officers and sailors
Armament	9 16-inch guns in 3 turrets 20 5-inch guns 80 40-millimeter guns (quad mounts) 49 20-millimeter
Builder	New York Naval Shipyard
Laid down	June 27, 1940
Launched	August 27, 1942
Commissioned	February 22, 1943
Decommissioned	March 24, 1949
Recommissioned	August 25, 1951
Decommissioned	February 24, 1958
Recommissioned	April 28, 1984
Decommissioned	October 26, 1990
Stricken	January 12, 1995
Preserved by	Pacific Battleship Center, Los Angeles, CA
Preservation Donations	www.pacificbattleship.com
Coordinates	33.742168, -118.277303

An inert sixteen-inch/50-caliber armor-piercing shell displayed with six replica powder bags is shown on the aft deck. The explosive force of all that powder could propel one of the shells twenty-four miles. *Nicholas A. Veronico*

exists today, more than seventy years after the United States was drawn into World War II.

American shipyards built a series of emergency cargo ships, the first known as Liberty ships (modeled after the British *Ocean*-class freighters) and the improved Victory ships. Eighteen shipyards built Liberties, and of those, six yards converted over to building Victory ships in 1943. In total, 2,710 Liberties and 531 Victories were built from 1943 to 1945. Today, three Liberty ships and three Victory ships have been restored and four of the six sail, with one Victory ship still under restoration with plans to have it sailing in the next couple of years.

The Liberty design was classified by the Maritime Commission as "EC2-S-C1"— "EC" for "emergency cargo" ship; "2" denoting the ship's size (waterline length from 400 to 450 feet; "S" for steam engine; and "C1" representing the ship design and modifications.

Each Liberty ship was 441 feet 6 inches long with a beam of 57 feet and a draft of 27 feet 9 inches. Liberty ships displaced 14,245 tons, of which approximately 8,500 tons was cargo. These ships were powered by a 2,500-horsepower triple expansion steam engine and could cruise at approximately eleven knots. The ships typically carried a crew of thirty-seven plus twenty-six Navy sailors, known as "Armed Guard," to man the ship's defensive guns (three three-inch/fifty-caliber, one five-inch/thirty-eight-caliber, and eight twenty-millimeter antiaircraft defense guns).

The ship's hulls were welded rather than riveted, and using assembly line techniques, it eventually took only forty-two days to build a Liberty ship. The record for building a Liberty ship was set by Henry J. Kaiser's Richmond shipyard in a

Although only five Victory ships are visible in this July 2002 photo, there are actually nine in this nest, alternating bow to stern. These are the last remaining Victory ships in the reserve fleet at Suisun Bay, California. The Victory ships are, in alphabetical order: *Barnard Victory*, *Hannibal Victory*, *Earlham Victory*, *Occidental Victory*, *Pan American Victory*, *Queens Victory*, *Rider Victory*, *Sioux Falls Victory*, and the *Winthrop Victory*. The stern of *LST-1158 Tioga County* is seen in the foreground. *Tioga County* was sunk by the US Navy during a SINKEX (sink at sea live-fire training exercise). When this photo was taken, only thirteen Victory ships remained in storage—nine in Suisun Bay, and two each at James River, Virginia, and Beaumont, Texas. *Nicholas A. Veronico*

Appearing rusted, their paint peeling, three of the nine Victory ships in the Suisun Bay mothball fleet are visible. The ship in the middle, stern to the camera with the solid white stack, is *Pan American Victory*, built at Permanente Metals Corp. (Kaiser Richmond) Yard No. 2, and launched on April 14, 1945. This ship was towed to the breakers in Brownsville, Texas, on November 30, 2009. *Nicholas A. Veronico*

pre-staged publicity stunt that saw *Robert E. Peary* constructed from keel laying to launch in only four days, 15.5 hours.

Of the two surviving Liberty ships in the United States, *John W. Brown* is berthed on the East Coast at Baltimore Harbor, Maryland, and *Jeremiah O'Brien* is tied up at San Francisco, California. *O'Brien*, the first Liberty ship to be rescued from the mothball fleets, is named for Capt. Jeremiah O'Brien (1744–1818), who captured the British schooner HMS *Margaretta* on June 12, 1775, at Machias, Maine, during the Revolutionary War. *Jeremiah O'Brien* was constructed at the New England Shipbuilding Corporation's South Portland, Maine, yard, and was launched on June 19, 1943.

On D-Day Plus Three, June 10, 1944, *Jeremiah O'Brien* delivered supplies to Allied troops flooding the beaches of France and taking the fight to Nazi-occupied Germany. Her delivery was made to Omaha Beach in

Interior of the restored wheelhouse of *Jeremiah O'Brien*. It was from here that the ship was guided from San Francisco Bay to the beaches of Normandy, France, for the fiftieth anniversary of D-Day. *Nicholas A. Veronico*

Normandy, the site of the American landings on June 6. *O'Brien* made a total of eleven trips from England to the beaches of France during the opening months of the invasion.

After overhaul, *O'Brien* made a trip to South America. In January 1945, she loaded ordnance for the island-hopping campaign against Japan and set sail for the Pacific Ocean. During the first half of 1945, *O'Brien* was supporting Allied troops in New Guinea and the Philippines. On June 25, 1945, she pointed her bow eastward for California. After a short overhaul, *O'Brien* loaded cargo in Southern California's San Pedro Harbor before sailing to India. From there, she delivered cargo to Allied troops in China and then called at Fremantle, Australia. From Australia, *O'Brien* returned to San Francisco. After her cargo was unloaded, the ship was sent to the mothball fleet in Suisun Bay, arriving on February 7, 1946.

In 1962, there were more than three hundred Liberty ships in storage at Suisun Bay. Admiral Tom Patterson was part of the team that surveyed the Liberties with the intent of selling them off for scrap. Patterson knew that *O'Brien* was a special ship and did everything he could to prevent the vessel from being scrapped. *O'Brien* was special because she was unaltered; everything looked the way it did when the ship served during World War II.

Recognized for being the last unaltered Liberty ship afloat, *Jeremiah O'Brien* is open to visitors at San Francisco's Pier Forty-Five. She makes a number of cruises about the bay each year. The pin-up "Jerry O'Brien" can be seen on the bow gun tub.
Nicholas A. Veronico

In 1978, the National Liberty Ship Memorial was established as a nonprofit corporation to acquire, preserve, and operate *Jeremiah O'Brien* as a floating monument to the men and women who built the ships as well as to the merchant seamen and Armed Guard sailors who served on board—many making the ultimate sacrifice.

Beginning in the summer of 1979, *O'Brien* volunteers worked for three months to prepare the Liberty ship to steam from the mothball fleet to her new home in San Francisco to begin her new career as an educational foundation and memorial. On October 6, 1979, *Jeremiah O'Brien* sailed under her own power from Suisun Bay, making the dreams of many a reality.

In 1994, *Jeremiah O'Brien* returned to England and France for the fiftieth anniversary commemoration of the June 6, 1944, D-Day landings. *O'Brien* was the largest ship from the June 1944 invasion of the European continent to return fifty years later. During her return stateside, *Jeremiah O'Brien* met the only other operating Liberty ship, *John W. Brown*, at sea near Cape Cod, marking the last time two of the war emergency ships crossed paths.

Today, *Jeremiah O'Brien* is seen by thousands each year as she is moored at Pier 45 at Fisherman's Wharf along San Francisco's waterfront. Her engine is run once a month, and that is a sight to see. If you've seen *Titanic* on the silver screen, then you've seen *Jeremiah O'Brien*'s engines at work as they were filmed for the movie. When the piston rods move to turn *Titanic*'s propellers, those are *O'Brien*'s with a little computer-generated magic to make them appear even larger than they are in real life. The Liberty ship also makes a number of cruises each year and can be seen on the San Francisco Bay sailing during the city's fleet week celebrations every October.

John W. Brown was built at the Bethlehem-Fairfield Shipyard in Baltimore, Maryland, sliding down the ways on Labor Day, September 7, 1942. The ship is

SPECIFICATIONS: LIBERTY SHIPS

Length, overall	441 feet 6 inches	
Beam	56 feet 10.75 inches	
Draft	27 feet 9.25 inches	
Displacement	10,920 tons	
Powerplant	Triple-expansion steam engine of 2,500 hp driving a single propeller	
Crew	36 merchant and 26 naval armed guard	
Armament	1 5-inch/38-caliber (stern) 1 3-inch/50-caliber (bow) 8 20-millimeter	
	John W. Brown	**Jeremiah O'Brien**
Maritime Commission Emergency Hull No.	312	806
Builder	Bethlehem-Fairfield Shipyard	New England Shipbuilding Corp.
Launched	September 7, 1942	June 18, 1943
Decommissioned	July 1983	February 7, 1946
Stricken	1998	1979
Preserved by	Project Liberty Ship	National Liberty Ship Memorial
Website	www.ssjohnwbrown.org	www.ssjeremiahobrien.org
Coordinates	39.267870, -76.569886	37.810992, -122.417955

The restored Liberty Ship *John W. Brown* sails through Buzzards Bay on approach to the Massachusetts Maritime Academy in Bourne on the Cape Cod Canal. When not underway, the ship can be toured at her home berth in Baltimore Harbor, Maryland. *Joan Burke/ Project Liberty Ship*

named for labor leader John W. Brown (1867–1941), whose last position was with the Industrial Union of Marine and Shipbuilding Workers of America, Local 4, in Maine. During the war, *John W. Brown* delivered cargo to the Soviet Union through the Persian Gulf and to ports in the Mediterranean Sea. The ship supported such major operations as the Anzio Landings and the invasion of Southern France.

Shortly after the end of World War II, *John W. Brown* was transferred to the City of New York, where she served as a floating maritime trade school. For thirty-six years the Liberty ship taught students how to prepare for careers in the merchant marine while maintaining the ship in working condition. The group Project Liberty

Ship was formed in 1978 as *John W. Brown* was nearing the end of its school career. The nonprofit obtained a commitment from MARAD to donate the ship in 1982, but they had problems developing an organization to support the ship based in the New York area. The Liberty ship sat idle for one year, and with no supporting organization on the horizon, MARAD moved the ship to the James River Reserve Fleet in Virginia. In 1987, Project Liberty Ship approached the Baltimore Museum of Industry, and with the museum's support, the effort began to move *John W. Brown* to Baltimore Harbor.

In 1990 and early 1991, the ship was dry-docked to address seaworthiness issues before she could become a museum. After a tremendous volunteer effort, the ship's engines were restarted and she made her first trip as a museum ship in August 1991. Since that date, volunteers have worked to return *John W. Brown* to her original configuration and to host "Living History" cruises.

The third surviving Liberty ship is *Hellas Liberty*, now on display in Piraeus Port, Athens, Greece. This ship began life as *Arthur M. Huddell*, and was launched at the St. Johns River Shipbuilding Co., Jacksonville, Florida, on December 7, 1943. During the war, *Arthur M. Huddell* was converted to a pipe transport, having holds four and five joined to enable the ship to carry seventy miles of pipe in one load. The ship was stored in the mothball fleet at Suisun Bay, California, from 1946 to 1956, when it was removed from storage and converted into a cable layer. This Liberty ship then supported the construction of the Distant Early Warning (DEW) Line (1957–1964), and then the underwater submarine detection Sound Surveillance System (SOSUS) network (1965–1984). *Arthur M. Huddell* was then laid up at the James River Reserve Fleet, where she donated parts to restore *John W. Brown*. In 2008, the ship was sold to Greece. After two years of restoration, she was renamed *Hellas Liberty* and placed on display.

PRESERVED VICTORY SHIPS

Two of the three surviving Victory ships are operational—*American Victory* and *Lane Victory*—and the third, *Red Oak Victory*, is completing its restoration.

Based in Tampa Bay, Florida, the American Victory Mariners Memorial and Museum is home to the operational *American Victory*, a ship that served in three wars. The ship's keel was laid down on March 30, 1945, at the California Shipbuilding Corp. (known as Calship) in Wilmington (Los Angeles Harbor). She was launched on May 24 and delivered on June 20. In what was later recognized as the closing days of the war, *American Victory* made her first voyage to the Philippines and China. Her route back to the United States took her to India and Egypt before calling at New York Harbor. She was then off to South America before making a European cruise during the second half of 1946 and 1947. In December 1947, she was laid up in the Hudson River Reserve Fleet until her next call to duty.

With the Korean War underway, *American Victory* was recalled to service on February 15, 1951. Her service was short, as she was laid up again on January 6, 1954, this time at Beaumont, Texas. A dozen years later, America needed additional cargo-carrying capacity to support troops engaged in Vietnam. *American Victory* was recalled on July 19, 1966. She sailed around the world, but was back in moth-balls at the James River Reserve Fleet by October 1969.

In 1985, MARAD activated her for a test to determine what it would take to put a forty-year-old ship back into service. After the test was completed, *American Victory* was maintained in a ready reserve status and could be removed from storage and sailing within 30 days. That state of readiness lasted three years, and a decade later, in 1998, the effort began to save the ship.

Captain John C. Timmel organized a group within the Tampa Bay community, and Victory Ship Inc. was formed to preserve *American Victory*. The ship arrived

The restored *Lane Victory* is berthed in Los Angeles Harbor, and benefits from its proximity to the TV and movie-making industry in Southern California. The ship has appeared in dozens of productions. *Lane Victory/Christopher Scott*

in Tampa Bay in September 1999, and work immediately began to make the ship presentable. The following year, *American Victory* moved to her permanent berth alongside The Florida Aquarium. The ship made a number of cruises from 2000 to 2009, and is now seeking donations for a planned dry-dock period.

American Victory's sister ship is *Lane Victory*, which was also constructed at Calship in Wilmington—*American Victory* is Calship Hull No. 792 and *Lane Victory* is Hull No. 794, with *Lane Victory* following about two weeks behind its sister ship. *Lane Victory* was laid down on April 5, 1945, launched on May 31, and delivered on June 27. She was able to get one cruise in before the war ended, delivering cargo to the Admiralty Islands. During her return trip, the Japanese formally surrendered and *Lane Victory* prepared for a second island-hopping trip. From October 8, 1945, to February 27, 1946, the ship hauled cargo to Guam, Saipan, Peleliu, and Oahu, before returning to San Francisco. Subsequent trips saw the ship making stops in Karachi, Pakistan, multiple locations in the Philippines, Hong Kong, Korea, and Japan before returning to San Francisco. *Lane Victory* then made a transit through the Panama Canal to cruise the Atlantic Ocean with ports of call on the US East

Red Oak Victory takes shape at Permanente Metals Corporation's Yard No. 1 in Richmond, California. The ship was launched on November 9, 1944, and delivered on December 5, 1944. For the remainder of World War II, *Red Oak Victory* served as an ammunition supply ship providing fourteen- and sixteen-inch shells to battleships in the Pacific Theater. *Richmond Museum of History*

Sponsor Edna Ray Reiley breaks a bottle of champagne across the bow of *Red Oak Victory* on November 9, 1945, while matron of honor Orpha Berens protects her eyes from the flying liquid. Moments later, *Red Oak Victory* slid backwards into the Richmond Inner Harbor Basin. She was fitted out and then delivered on December 5, 1944. The ship loaded ammunition at northern California's Port Chicago Naval Magazine and then sailed for Ulithi Atoll in the Caroline Islands to begin her war service. *Richmond Museum of History*

Coast and in Europe. Traveling to various ports in the Mediterranean Sea, the ship exited the Suez Canal headed for Japan. She eventually returned to San Francisco in March 1948. Her World War II–era travels had come to an end, and *Lane Victory* was tied up in the ship nests at the Suisun Bay reserve fleet.

Called up for the Korean War, *Lane Victory* was back in action on October 10, 1950, sailing from the port at the Oakland Army Base, California, to Yokohama, Japan. For the next three years, *Lane Victory* supported military operations in Korea; however, she did make one Atlantic crossing in early 1952.

Lane Victory was back in the Suisun Bay reserve fleet from October 1953 until returned to service for the Vietnam War in August 1966. The Department of Defense recalled 172 Victory ships for the buildup in Southeast Asia, and *Lane Victory* was one of them. Cargoes were hauled to Guam, the Philippines, Japan, and Vietnam to support military forces in the region. Her tasks complete, *Lane Victory* was returned to storage at Suisun Bay on April 29, 1970.

In 1987, World War II merchant marine veterans Joe Vernick and John Smith began looking for a ship suitable for the US Merchant Marine Veterans of World

Red Oak Victory was returned to her birthplace at the Port of Richmond on San Francisco Bay and is seen here in October 2004, shortly after she emerged from the mothball fleet in nearby Suisun Bay. The ship shows the ravages of years of exposure without regular maintenance during its storage. *Nicholas A. Veronico*

On May 20, 2014, *Jeremiah O'Brien*, right, steamed past *Red Oak Victory*, tied up at Richmond Harbor. Soon the ships will be sailing together as the restoration of the *Red Oak Victory* is progressing nicely. *Roger Cain*

War II, a nonprofit, educational institution. The pair found *Lane Victory* in Suisun Bay in 1987, and after three years of dedicated work, the ship was towed into Los Angeles Harbor on June 12, 1989, to begin restoration. In the ensuing years the ship was restored, and it now makes a number of cruises each year. In addition, *Lane Victory* has starred in a number of TV and Hollywood films.

The third surviving Victory is *Red Oak Victory*, the 558th ship built at the Kaiser Shipyards in Richmond, California. The ship is named for the Iowa town that lost twenty-seven men from Company M, Iowa National Guard, during the Battle for Kasserine Pass, Tunisia, in February 1943. By war's end, fifty men from Red Oak, population 5,600, lost their lives, which was the highest per-capita loss of any US city.

Red Oak Victory was launched on November 9, 1944, and delivered on December 5, 1944. For her initial cruise, the ship loaded cargo and ammunition in the San Francisco Bay Area, with fourteen- and sixteen-inch cannon shells being taken on at Port Chicago Naval Magazine. From there, she sailed to Ulithi Atoll in the Caroline Islands (located between Papua New Guinea and Guam) to issue ammunition to US Navy ships staging through the atoll for the attack on Japan.

In May 1946, *Red Oak Victory* was withdrawn from service and stored in the reserve fleet nests at Suisun Bay. The ship was recalled for duty for the Korean War, then for a humanitarian mission to haul grain to India and Pakistan in 1956. She saw service in support of operations in Vietnam from 1965 to December 1968, when *Red Oak Victory* was retired for the final time. She made her home again at Suisun Bay.

When searching for a ship to represent the City of Richmond, California, where 747 ships were built for World War II, the Richmond Museum of History, headed by Lois Boyle, surveyed *Red Oak Victory*. The Suisun Bay reserve fleet is less than thirty miles from the Port of Richmond, and its acquisition and display would make a nice complement to the National Park Service's "Rosie the Riveter World War II Home Front Park," then being installed along the city's waterfront at the site of the former Kaiser shipyards.

By 1997, the ship was assigned to the Richmond Museum of History, whose team began working on the ship while it was still tied up at Suisun Bay. One year later, on September 17, 1998, *Red Oak Victory* was towed from the mothball fleet to the Richmond Harbor to begin its full restoration. Externally the ship looks very presentable, while the interior is receiving attention from volunteers. The ship's engine should be lit soon, and in the not too distant future, this ship will begin San Francisco Bay cruises. As the restoration progresses, the Richmond Museum Association presents before and after photos on its website (www.ssredoakvictory.com).

The World War II–era ships hidden in the mothball fleet nests that have survived are few, yet the work to restore and maintain them is enormous and requires a great deal of dedication by large numbers of volunteers and lots of money. Support them; they're all owed a tip of the cap for preserving our World War II maritime heritage.

SPECIFICATIONS: VICTORY SHIP

Length, overall	455 feet 3 inches
Beam	62 feet
Draft	28 feet
Displacement	15,200 tons
Powerplants	Steam turbine engine of 6,000 hp at 90 to 100 rpm driving a single 19-foot-diameter propeller
Crew	62 merchant and 28 naval armed guard
Armament	1 5-inch/38-caliber dual purpose 1 3-inch/50-caliber 1 40-millimeter 8 20-millimeter 2 .50-caliber machine guns *American Victory* *Lane Victory* *Red Oak Victory*

Ship name	*American Victory*	*Lane Victory*	*Red Oak Victory*
Maritime Commission Hull No.	792	794	544
Builder	California Shipbuilding Corp.	California Shipbuilding Corp.	Permanente Metals Corp. No. 1
Launched	May 24, 1945	May 31, 1945	November 9, 1944
Decommissioned	January 6, 1954	April 29, 1970	December 1968
Stricken	April 1999	March 1988	October 1996
Preserved by	American Victory Ship Mariners Memorial Museum	SS *Lane Victory*	Richmond Museum of History
Website	americanvictory.org	lanevictory.org	ssredoakvictory.com
Coordinates	27.943645, -82.444120	33.717288, -118.273148	37.904823, -122.364414

EPILOGUE

SEE A WORLD WAR II
SHIPWRECK FOR YOURSELF

FOR THOSE WILLING TO GET WET, there are a great number of World War II–era shipwrecks to dive the world over. Depending upon one's technical skill, US East Coast divers can explore everything from tugs to freighters to U-boats. Those willing to venture to World War II's former battlefields will also find a great number of shipwrecks, most available through guided diving tours.

In the Florida panhandle waters of the Gulf of Mexico, divers of all skill levels, from snorkelers to professionals, can enjoy the recently created Florida Panhandle Shipwreck Trail, which was officially announced on June 22, 2012. The Florida Department of State's Underwater Archaeology Team developed a series of dive sites that stretch along the coast from Pensacola to Destin to Panama City and Port St. Joe.

There are twelve dive sites ranging from the aircraft carrier USS *Oriskany* (CV-34), an *Essex*-class carrier that was constructed during World War II that lies between 80 and 212 feet below the surface, to the World War II *Admirable*-class minesweeper USS *Strength* (AM-309), which survived a midget submarine's torpedo attack and a kamikaze attack that was thwarted when *Strength*'s gunners shot it out of the sky, to *Vamar*, one of Admiral Byrd's ships that participated in his 1928 Antarctic Expedition and was sunk on March 21, 1942, in just twenty-five feet of water.

In addition, the Florida Panhandle Shipwreck Trail is supported by dive shops in the area and has an interactive website with information and videos about all the sunken ships.

WANT TO SEE A WORLD WAR II–ERA
SHIPWRECK, BUT YOU'RE NOT A DIVER?

When it's vacation time and you're heading west, a stop on Oahu and a visit to the World War II Valor in the Pacific National Monument is in order. The USS *Arizona* Memorial and its visitors center are a great place to start. Taking the boat tour to

The aircraft carrier *Oriskany* (CV-34) was sunk twenty-four miles off the coast of Pensacola, Florida, on May 17, 2006, to form an artificial reef. After twenty-five years of service to the Navy in operations in Korea, Vietnam, and the Mediterranean and eight years on the bottom of the Gulf of Mexico, *Oriskany* is providing a recreation diving destination off the coast of the Florida panhandle. *Photographer's Mate 2nd Class Jeffrey P. Kraus/US Navy*

It took the 888-foot-long *Oriskany* about thirty-seven minutes to sink below the surface. *Photographer's Mate 2nd Class Jeffrey P. Kraus/US Navy*

The largest ship on the Florida Shipwreck Trail is the aircraft carrier *Oriskany*. The carrier's depth ranges from 80 to 212 feet below the surface. In 2014, *Oriskany* was named one of the top twenty-five dive sites in the United States by *Scuba Diving* magazine. *Jim Phillips/Florida Department of State, Division of Historical Resources, Bureau of Archaeological Research*

MULTIBEAM SONAR IMAGE: ANA GARCIA-GARCIA AND MATTHEW LEVEY

There are twelve shipwrecks along the Florida Panhandle Shipwreck Trail, and dive shops and charter boats will validate a diver's "passport" with a sticker for each wreck dived. One of the twelve dive sites is the *Strength* (AM-309), a World War II minesweeper that survived both a midget submarine and a kamikaze attack, seen in a side-scan sonar image. *Ana Garcia and Matthew Levey/Florida Department of State, Division of Historical Resources, Bureau of Archaeological Research*

see the *Arizona* Memorial is a must-see. On the opposite side of Ford Island lies the target ship *Utah*, although access to this location is restricted.

Associated with the Valor in the Pacific National Monument is the Battleship *Missouri* Memorial, berthed facing the *Arizona* Memorial. The contrast in the size of the two ships is most evident in aerial photos, and it is extremely moving to walk on the *Arizona* Memorial—the site where the war began—and then tour *Missouri* and stand in the exact location where Gen. Douglas MacArthur accepted the surrender of the Japanese.

Located adjacent to the Pearl Harbor Visitor Center is the USS *Bowfin* Submarine Museum and Park, home to the restored *Balao*-class submarine and a fantastic interpretive center. USS *Bowfin* (SS-287) was launched at the Portsmouth Navy Yard, Maine, on December 7, 1942, one year after the war began, and was nicknamed the "Pearl Harbor Avenger." Bowfin Park also includes a memorial honoring all fifty-two American submarines and their 3,500 crewmen lost during the war.

The remains of Landing Ship Tank 480 (*LST-480*) stand as a reminder of the more than 130 men and six ships lost during the May 21, 1944, West Loch Disaster. At the time, twenty-nine LSTs were loading stores and equipment for the forthcoming invasion of the Mariana and Palau Islands, known as Operation Forager. It is believed a 110-millimeter mortar shell exploded aboard *LST-353*, which then set off barrels of gasoline stored nearby. As the conflagration grew, five other LSTs, all in various stages of loading, became involved. Many of the LSTs were able to move to safer waters, either under their own power or with the assistance of tugs. The LSTs destroyed during the disaster were *LST-39*, *LST-43*, *LST-69*, *LST-179*, *LST-353*, and *LST-480*. Along with the LSTs, seventeen tracked landing vehicles (LVTs) were destroyed.

The LSTs and the LVTs were cleared from West Loch and their remains dumped at sea. Information about the disaster was blacked out during the war, and was finally declassified in 1960. Today, only the bow of *LST-480* is visible above the waters of Pearl Harbor.

For those a bit more adventurous, Atlantis Submarines offers the opportunity to see a shipwreck approximately a hundred feet below the surface without ever getting wet. The company acquired the 1940s-vintage yard oiler *YO-257* in the late 1980s when it was sold surplus. Atlantis cleaned the ship, loaded it with gravel so it would land upright on the bottom, and scuttled the ship in the waters off Waikiki.

YO-257 is 165 feet long with a thirty-foot beam and sits at the bottom at 110 feet. The oiler's top deck is eighty feet below the surface and the smokestack reaches up to sixty-five feet. Submarine passengers are able to view the wreck in very clear water and to see a large variety of marine life ranging from eels and turtles to reef fish and sharks.

Water-level view of the chaos that occurred as twenty-nine LSTs were loading in Pearl Harbor's West Loch for the invasion of the Mariana and Palau Islands. Six LSTs were consumed in the explosion and resulting fires. *US Navy*

Aerial view of *LST-480* as crews try to extinguish the fully loaded ship. The aft section of the ship is completely burned-out. *US Navy*

SHIPWRECKS IN THE WILD

The former World War II battlefields in the Aleutian Islands are a shipwreck hunter's paradise. On June 6, 1942, a Japanese force five hundred strong invaded the island of Kiska. Four months later, on October 14, 1942, the 385-foot-long freighter *Borneo Maru* arrived in Gertrude Cove, Kiska Island, to offload supplies to the Japanese garrison. The next morning, at 9:00 a.m., more than twenty US Army Air Forces aircraft bombed the ship. Although none were direct hits, the concussion from several near misses split open the ship's hull. *Borneo*

The rusted hull of USS *LST-480* remains in the water at West Loch, Joint Base Pearl Harbor–Hickam. The remains honor the 163 soldiers, sailors, and Marines who died and the 396 who were wounded when a fuel and ammunition explosion occurred at West Loch during World War II on May 21, 1944. *Library of Congress and Mass Communication Spc. Seaman Johans Chavarro/US Navy*

Maru tried to get underway but was subsequently beached and abandoned. The remainder of the ship's cargo was offloaded while American aircraft continued to bomb the ship. In the seventy years since the ship went aground, it has slowly deteriorated, its hull caving in as it slides further and further beneath the waves.

SPECIFICATIONS:
LST 480 (LANDING SHIP TANK)

Length, overall	328 feet
Beam	50 feet
Draft	14 feet 1 inch (maximum navigation)
Displacement	4,080 tons with 1,675 ton load
Powerplant	2 900 hp General Motors 12-567A Diesel engines driving twin screws with dual rudders
Crew	13 officers, 104 sailors
Armament	2 twin 40mm; 4 single 40mm; 12 single 20mm cannon
Builder	Kaiser Cargo
Launched	October 29, 1942
Commissioned	May 3, 1943
Sunk	May 21, 1944
Location	West Loch, Pearl Harbor, Hawaii
Coordinates	21.357266, -157.997221

SPECIFICATIONS:
BORNEO MARU (PASSENGER/ CARGO SHIP, IMPERIAL JAPANESE ARMY)

Length	385 feet
Beam	51 feet
Draft	11 feet
Displacement	5,863 tons
Powerplants	1 triple-expansion steam engine, 440 hp
Speed	12.5 knots cruising
Builder	Kawasaki Shipbuilding, Kobe, Japan
Launched	October 28, 1917 (as *Daifuku Maru No. 14*)
Sunk	October 21, 1942, by US Army Air Forces aircraft
Location	Gertrude Cove, Kiska Island, Alaska
Coordinates	51.934631, 177.456124

If you'd like to combine a snorkel trip with an up-close tour of some World War II–vintage shipwrecks, the Ocean Riders' circumnavigation of Lanai is highly recommended. The company uses thirty-foot rigid-hull inflatable boats certified to carry twenty-two passengers, but only eighteen are taken. The seven-hour trip covers seventy miles and departs from Lahaina, Maui. The boat crosses the Auau

Channel, and the tour starts along Lanai's north coast. The first site seen is the wreck of the World War II gasoline barge *YOGN-42*, which gives the area its name: "Shipwreck Beach."

YOGN-42 is a non-self-propelled gasoline barge, one of twenty-four steel-reinforced concrete ships built during World War II by Concrete Ship Constructors of National City, California. These gasoline barges were fitted with four twenty-millimeter antiaircraft guns, two mounted forward and two mounted aft, as well as a deckhouse at the stern that served as quarters for its thirteen crewmen. Launched on March 23, 1943, as *YOG-42*, it was commissioned into service on May 22, 1943 (later its designation was changed to *YOGN-42*).

On December 12, 1943, *YOGN-42* was servicing ships in the South Pacific area around Espiritu Santo. The gasoline barge was under tow 150 miles from Espiritu Santo by USS *Navajo* (AT-64) when the two were attacked by the Japanese submarine *I-39*. Hit by a torpedo, *Navajo* was sunk. *YOGN-42* eventually returned to Pearl Harbor, where it ended its naval career in May 1946. The barge reportedly ended up on the beach when its tow broke during a storm in 1949. The ship sits hard on a reef with its port side to the weather as waves continually break over its bow, making *YOGN-42* look as if it is underway.

The wreck of the Japanese transport *Borneo Maru* is deteriorating quickly. The ship was damaged while unloading supplies during the invasion of Kiska on October 15, 1942; in the following days, the transport was repeatedly bombed by American aircraft and was abandoned in Kiska Island's Gertrude Cove. The sea is rapidly reclaiming the ship's remains. *Caroline Funk*

The World War II gasoline barge *YOGN-42* is hard on the reef on the north coast of the island of Lanai. This barge saw combat in the South Pacific and is said to have ended up on the beach while being towed during a storm in 1949. The eight-mile stretch of beach where *YOGN-42* rests is named Shipwreck Beach. *Nicholas A. Veronico*

Depending upon the tide conditions, the remnants of six other shipwrecks of various vintages can be seen along this eight-mile stretch of beach as the boat travels west. The last of the wrecks is the yard oiler *YO-21*, which sits in the surf line, waves slowly taking their toll on the ship. *YO-21* was a 160-foot-long, single-screw, steel hull tanker that was built in 1918 by the Tank Shipbuilding Co. of Newberg, New York. The oiler entered service with the US Navy in 1924, and was on duty at Pearl Harbor during the December 7, 1941, attack. After the war, the ship was stricken from the Navy's inventory, and the locals say the ship was dumped on the north shore of Lanai with numerous other surplus vessels.

For those who are not scuba divers, there are a number of opportunities to see the ships and shipwrecks of World War II. Depending on your location and your level of adventure, these warships can be seen from shore, ship, or submarine.

At the western end of Lanai's Shipwreck Beach is *YO-21*, an oiler built at the end of World War I. The ship was present at Pearl Harbor during the Japanese attack on December 7, 1941. The ship sits bow-on to the shore and is starting to cave in on itself.
Nicholas A. Veronico

SPECIFICATIONS: *YOGN-42*
(NON-SELF-PROPELLED GASOLINE BARGE)

Length	375 feet
Beam	56 feet
Draft	26 feet 6 inches
Displacement	5,410 tons; 6,600 tons fully loaded
Powerplant	none
Crew	13
Armament	4 20-millimeter antiaircraft guns
Builder	Concrete Ship Construction
Launched	March 23, 1943
Commissioned	May 1943
Sunk	late 1940s
Location	North shore, Lanai Island, Hawaii
Coordinates	20.921196, -156.910062

SPECIFICATIONS: *YO-21*
(SELF-PROPELLED FUEL OIL BARGE)

Length	161 feet
Beam	25 feet 1 inch
Draft	unknown
Displacement	335 tons
Powerplants	1 triple-expansion steam engine driving 1 propeller
Crew	20
Armament	none
Cargo Capacity	3,563 barrels
Builder	Tank Ship Building, Co.
Launched	1918
Commissioned	1924
Decommissioned	August 22, 1946
Grounded	early 1950s
Location	North shore, Lanai Island, Hawaii
Coordinates	20.929305, -157.000712

APPENDIX I

US NAVY SHIP FORCE LEVELS, 1939–1945

	6/30/1939	6/30/1940	12/7/1941	12/31/1942	12/31/1943	12/31/1944	8/14/1945
Battleships	15	15	17	19	21	23	23
Carriers, Fleet	5	6	7	4	19	25	28
Carriers, Escort	0	0	1	12	25	65	71
Cruisers	36	37	37	39	48	61	72
Destroyers	127	185	171	224	332	367	377
Frigates	0	0	0	0	234	376	361
Submarines	58	64	112	133	172	230	232
Mine Warfare	29	36	135	323	551	614	586
Patrol	20	19	100	515	1,050	1,183	1,204
Auxiliary	104	116	210	392	564	993	1,267
Surface Warships	178	237	225	282	635	827	833
Total Active	380	394	478	790	1,782	3,699	6,768

September 1939: World War II begins in Europe when Germany and the USSR invade Poland
May 8, 1945: World War II ends in Europe
August 14, 1945: V-J Day (August 15 in the Western Pacific)
September 2, 1945: Pacific War formally ends
Source: Naval Historical Center

APPENDIX II

PRESERVED WORLD WAR II SHIPS

Vessel number	Name	Date	Location
Battleships			
BB-39	*Arizona*	1915	USS *Arizona* Memorial, Pearl Harbor, Hawaii
BB-55	*North Carolina*	1941	USS *North Carolina* Battleship Memorial, Wilmington, North Carolina
BB-59	*Massachusetts*	1942	Battleship Cove, Fall River, Massachusetts
BB-60	*Alabama*	1942	Battleship Memorial Park, Mobile, Alabama
BB-61	*Iowa*	1942	Pacific Battleship Center, Los Angeles, California
BB-62	*New Jersey*	1943	Battleship *New Jersey* Museum and Memorial, Camden, New Jersey
BB-63	*Missouri*	1944	USS *Missouri* Memorial Association, Pearl Harbor, Hawaii
BB-64	*Wisconsin*	1943	Nauticus/Hampton Roads Naval Museum, Norfolk, Virginia
Aircraft Carriers			
CVS-10	*Yorktown*	1943	Patriot's Point Naval and Maritime Museum, Mount Pleasant, South Carolina
CVS-11	*Intrepid*	1943	Intrepid Sea-Air-Space Museum, New York City
CVS-12	*Hornet*	1943	USS *Hornet* Museum, Alameda, California
AVT-16	*Lexington*	1943	*Lexington* on the Bay Museum, Corpus Christi, Texas
CV-41	*Midway*	1945	USS *Midway* Museum, San Diego, California

Vessel number	Name	Date	Location
Cruisers			
CL-92	*Little Rock*	1944	Buffalo and Erie County Naval and Military Park, Buffalo, New York
Destroyers/Destroyer Escorts			
DD-537	*The Sullivans*	1943	Buffalo and Erie County Naval and Military Park, Buffalo, New York
DD-661	*Kidd*	1943	Louisiana Naval War Museum, Baton Rouge, Louisiana
DD-724	*Laffey*	1943	Patriot's Point Naval and Maritime Museum, Mount Pleasant, South Carolina
DD-793	*Cassin Young*	1943	Boston National Historical Park, Boston, Massachusetts
DD-850	*Joseph P. Kennedy Jr.*	1945	Battleship Cove, Fall River, Massachusetts
DD-886	*Orleck*	1945	SE Texas War Memorial and Heritage Foundation, Orange County, Texas
DE-238	*Stewart*	1943	Seawolf Park, Galveston, Texas
DE-766	*Slater*	1944	Destroyer Escort Historical Foundation, Albany, New York
Tank Landing Ships			
LST393		1943	Great Lakes Naval and Maritime Museum, Muskegon, Michigan
LST325		1942	Battleship Memorial Park, Mobile, Alabama
Submarines			
SS-224	*Cod*	1943	Cleveland Coordinating Committee, Cleveland, Ohio
SS-228	*Drum*	1941	Battleship Memorial Park, Mobile, Alabama
SS-236	*Silversides*	1941	Great Lakes Naval & Maritime Museum, Muskegon, Michigan
SS-244	*Cavalla*	1944	Seawolf Park, Galveston, Texas
SS-245	*Cobia*	1944	Wisconsin Maritime Museum, Manitowoc, Wisconsin
SS-246	*Croaker*	1944	Buffalo and Erie County Naval and Military Park, Buffalo, New York
SS-287	*Bowfin*	1943	Pacific Fleet Memorial Association, Pearl Harbor, Hawaii
SS-297	*Ling*	1945	New Jersey Naval Museum, Hackensack, New Jersey
SS-298	*Lionfish*	1944	Battleship Cove, Fall River, Massachusetts
SS-310	*Batfish*	1943	Muskogee War Memorial Park, Muskogee, Oklahoma

Vessel number	Name	Date	Location
SS-319	*Becuna*	1944	Independence Seaport Museum, Philadelphia, Pennsylvania
SS-343	*Clamagore*	1945	Patriot's Point Naval and Maritime Museum, Mount Pleasant, South Carolina
SS-383	*Pampanito*	1943	National Maritime Museum Association, San Francisco, California
SS-423	*Torsk*	1944	Baltimore Maritime Museum, Baltimore, Maryland
SS-481	*Requin*	1945	Carnegie Science Center, Pittsburgh, Pennsylvania

Minesweepers

AM-240	*Hazard*	1944	Military Historical Society, Freedom Park, Omaha, Nebraska

Patrol Boat

PT 305		1944	National World War II Museum, New Orleans, Louisiana
PT 309		1944	Admiral Nimitz Museum, Kemah, Texas
PT 617		1945	Battleship Cove, Fall River, Massachusetts
PT 658		1945	USN/Save the PT Boat Inc., Portland, Oregon
PT 724		1945	Liberty Aviation Museum, Port Clinton, Ohio
PT 728		1945	Liberty Aviation Museum, Port Clinton, Ohio
PT 796		1945	PT Boats Inc./Battleship Cove, Fall River, Massachusetts

Army Tugs

ST-695	*Angels Gate*	1944	Los Angeles Maritime Museum, San Pedro, California
LT-5		1944	H. Lee White Marine Museum, Oswego, New York

Vessel number	Name	Date	Location
------	*Marquette*	—	Great Lakes Naval and Maritime Museum, Chicago, Illinois
Landing Craft			
LSM-45		1944	Freedom Park/Omaha Military Historical Society, Omaha, Nebraska
LCI(L)-73		1944	Amphibious Forces Memorial Museum, Portland, Oregon
Liberty Ships			
—	*Jeremiah O'Brien*	1932	National Liberty Ship Memorial, San Francisco, California
—	*John Brown*	1942	Project Liberty Ship, Baltimore, Maryland
Victory Ships			
AK-235	*Red Oak Victory*	1944	Richmond Museum of History, Richmond, California
—	*American Victory*	1945	American Victory Mariners Memorial and Museum, Tampa, Florida
—	*Lane Victory*	1945	US Merchant Marine Veterans of World War II, San Pedro, California
German U-Boats			
U-505		1941	Museum of Science and Industry, Chicago, Illinois
U-534		1942	U-Boat Story, Birkenhead, England
U-995		1943	Marine Memorial, Laboe, Germany
U-2540		1945	German Maritime Museum, Bremerhaven, Germany
Japanese Midget Submarines			
—	*HA-8*	—	USN Submarine Force Museum, Groton, Connecticut
—	*HA-19*	—	Admiral Nimitz Museum, Fredericksburg, Texas

BIBLIOGRAPHY AND
SUGGESTED READING

BOOKS

Bailey, Dan E. *World War II Wrecks of the Truk Lagoon*. Redding, CA: North Valley Diver Publications, 2000.

_____. *WWII Wrecks of the Kwajalein and Truk Lagoons*. Redding, CA: North Valley Diver Publications, 1989.

_____. *World War II Wrecks of Palau*. Redding, CA: North Valley Diver Publications, 1991.

_____. Ballard, Robert D., with Rick Archbold. *The Lost Ships of Guadalcanal: Exploring the Ghost Fleet of the South Pacific*. Toronto: Madison Press Books, 1993.

Ballard, Robert D., and Rick Archbold. *Return to Midway: The quest to find the Yorktown and the other lost ships from the pivotal battle of the Pacific War*. Washington: National Geographic, 1999.

Beasant, John. *Stalin's Silver: The Sinking of the USS John Barry*. New York: St. Martin's Press, 1999.

Blair, Clay. *Hitler's U-Boat War: The Hunted, 1942–1945*. New York: Random House, 1998.

Boyd, Carl, and Akihiko Yoshida. *The Japanese Submarine Force and World War II*.

Buckley, Capt. Robert J. (USNR, retired) *At Close Quarters: PT Boats in the United States Navy*. Washington: Naval History Division, 1962.

Burlingame, Burl. *Advance Force Pearl Harbor*. Annapolis: Naval Institute Press. 1992.

Craddock, John. *First Shot: The Untold Story of the Japanese Minisubs That Attacked Pearl Harbor*. New York: McGraw-Hill, 2006.

Cressman, Robert J. *The Official Chronology of the U.S. Navy in World War II*. Annapolis: Naval Institute Press, 2000.

Geoghegan, John J. *Operation Storm: Japan's Top Secret Submarines and Its Plan to Change the Course of World War II*. New York: Broadway Books, 2013.

Jaffe, Capt. Walter W. *The Liberty Ships: From A (A. B. Hammond) to Z (Zona Gale)*. Palo Alto, CA: The Glencannon Press, 2004.

_____. *The Victory Ships: From A (Aberdeen Victory) to Z (Zanesville Victory)*. Palo Alto, CA: The Glencannon Press, 2006.

Kahn, David. *Seizing the Enigma: The Race to Break the German U-Boat Codes, 1939–1943*. New York: Barnes and Noble Books, 2001.

Lenihan, Daniel. *Submerged: Adventures of America's Most Elite Underwater Archeology Team*. New York: New Market Press, 2003.

Macdonald, Rod. *Force Z Shipwrecks of the South China Sea, HMS Prince of Wales and HMS Repulse.* Dunbeath, Caithness, Scotland: Whittles Publishing Ltd., 2013.

Miller, David. *U-Boats: The Illustrated History of the Raiders of the Deep.* Washington: Brassey's, 2000.

Newhart, Max R. *American Battleships: A Pictorial History of BB-1 to BB-71 with Prototypes Maine and Texas.* Missoula, MT: Pictorial Histories Publishing, 2001.

Niestlé, Axel. *German U-Boat Losses During World War II: Details of Destruction.* Annapolis: U.S. Naval Institute Press, 1998.

O'Kane, Richard H. *Wahoo: The Patrols of America's Most Famous World War II Submarine.* Novato, CA: Presidio Press, 1996.

Penrose, Barrie. *Stalin's Gold: The Story of HMS Edinburgh and its Treasure.* Boston: Little, Brown and Company, 1982.

Rohwer, Jürgen. *Axis Submarine Successes of World War Two: German, Italian and Japanese Submarine Successes, 1939–1945.* Annapolis: Naval Institute Press, 1999.

Sakaida, Henry, and Gary Nila. *I-400: Japan's Secret Aircraft Carrying Strike Submarine, Objective Panama Canal.* Crowborough, East Sussex, England: Hikoki Publications, 2006.

Scalia, Joseph Mark. *Germany's Last Mission to Japan: The Failed Voyage of U-234.* Annapolis: Naval Institute Press, 2000.

Sheard, Bradley. *Lost Voyages: Two Centuries of Shipwrecks in the Approaches to New York.* New York: Aqua Quest Publications, 1998.

Sharpe, Peter. *Beyond Sportdiving! Exploring the Deepwater Shipwrecks of the Atlantic.* Birmingham, AL: Menasha Ridge Press, 1996.

_____. *U-Boat Fact File: Detailed Service Histories of the Submarines Operated by the Kriegsmarine 1935–1945.* Earl Shilton, Leicester, England: Midland Counties Publications, 1998.

Stevens, Peter F. *Fatal Dive: Solving the World War II Mystery of the USS Grunion.* Washington: Regenery History, 2012.

Veronico, Nicholas A. *World War II Shipyards by the Bay.* Charleston, SC: Arcadia Publishing, 2007.

Veronico, Nicholas A. and Armand H. Veronico. *Battlestations! American Warships of World War II.* Osceola, Wisconsin: Motorbooks International, 2001.

Washichek, Richard J. *PT 658: An Illustrated Look into the World War II Higgins PT Boat and Its Restoration.* Portland, OR: Save the PT Boat Inc., 2013.

DOCUMENTS

Delgado, James P., Daniel J. Lenihan, and Larry E. Murphy. "The Archeology of the Atomic Bomb: A Submerged Cultural Resources Assessment of the Sunken Fleet of Operation Crossroads at Bikini and Kwajalein Atoll Lagoons, Republic of the Marshall Islands." Southwest Cultural Resources Center Professional Papers, No. 37. Santa Fe: National Park Service (Submerged Cultural Resources Unit), 1991.

Hay, S. M. "U.S.S. Peterson (DE-152) Anti-Submarine Action by Surface Ship," April 27, 1944.

Headquarters, Eastern Sea Frontier. "War Diary," April 1944.

Headquarters, Eastern Sea Frontier. "Enemy Action and Distress Diary; Activity Within Eastern Sea Frontier," April 16, 1944.

Jeffery, Bill (Chuuk Historic Preservation Office). "War in Paradise: World War II Sites in Truk Lagoon." Chuuk, Federated States of Micronesia, 2003.

Lenihan, Daniel J., ed. "Submerged Cultural Resources Study: USS *Arizona* Memorial and Pearl Harbor National Historic Landmark." Southwest Cultural Resources Center Professional Papers, No. 23. Santa Fe: National Park Service (Submerged Cultural Resources Unit), 1989.

Office of the Historian, Joint Task Force One. "Operation Crossroads: The Official Pictorial Record." New York: Wm. H. Wise and Co., 1946.

Sessions, W. A. "U.S.S. Gandy (DE-764) War Diary for the month of April 1944."

War Diary. "Composite Squadron FORTY-TWO (VC-42), April 15, 1943 to September 24, 1944."

War Diary. "U.S.S. Croatan March 24, 1944 to May 11, 1944."

War Diary. "U.S.S. Joyce (DE-317) Month of April 1944."

War Patrol, "U.S.S. Grunion (SS-216) April 11, 1942 to June 30, 1942."

PERIODICALS

Doyle, Greg. "HMS Repulse: Diving into British Naval History." *Advanced Diver Magazine.*
Online at: www.advanceddivermagazine.com/articles/repulse/repulse.html.

Lambert, Chip, and Dan Bailey. "The Wreck That Almost Sank a President." *Wreck Diving Magazine.* Issue 30.

INTERNET RESOURCES

American Merchant Marine at War
www.usmm.org

The website for information on the US Merchant Marine as well as Liberty and Victory ships, a shipmate search feature, and resources for genealogists and veterans.

Battle of the Atlantic Museum
www.thebattleofatlanticmuseum.ca

Website for George Cruickshank's history of the Battle of the Atlantic, and a showcase of his private museum of artifacts from both sides of the conflict. During the winter months, his collection is on display at HMCS Naden Naval and Military Museum in Esquimalt, British Columbia, Canada. Be sure to visit his *U-889* Project webpages, where he details the construction of a full-size replica of *U-889*'s conning tower.

The Bent Prop Project
www.bentprop.org

Patrick Scannon, MD, PhD, has been traveling to the Palau Islands to search for those missing in action from both sides of the conflict. Follow his outstanding work online.

Combined Fleet (Imperial Japanese Navy)
www.combinedfleet.com

Detailed information and photos of everything Japanese Navy—ships, torpedoes, guns, and war production statistics.

Florida Panhandle Shipwreck Trail
www.floridapanhandledivetrail.com

Twelve historic shipwrecks along the Florida Panhandle with excellent information on each of the dive destinations.

J-Aircraft
www.j-aircraft.com

Research, photos, and message board covering Japanese aircraft, ships, and historical research.

Lust4Rust
www.petemesley.com

Pete Mesley's website for wreck diving instruction and diving trips to Bikini Atoll, Truk Lagoon, and other historic sites.

Rod MacDonald—Dive Guide, Author, Speaker
www.rod-macdonald.co.uk

Diver Rod MacDonald has traveled the world over in search of shipwrecks, from HMS *Prince of Wales* and HMS *Repulse* to Truk Lagoon to Scapa Flow, he's been there. MacDonald has documented his explorations in a series of books written for the diver and the history enthusiast.

National Oceanic and Atmospheric Administration
National Marine Sanctuaries
sanctuaries.noaa.gov

The Office of National Marine Sanctuaries serves as the trustee for a network of fourteen protected marine areas encompassing more than 170,000 square miles of marine and Great Lakes waters from Washington State to the Florida Keys and from Lake Huron to American Samoa. The network includes a system of thirteen national marine sanctuaries and the Papahānaumokuākea Marine National Monument (Hawaiian Islands). In addition, the Sanctuaries program is the steward of numerous historic shipwrecks detailed on these pages. Be sure to check out the Sanctuaries' Maritime Heritage webpages and downloadable reports.

National Park Service • World War II Valor in the Pacific
www.nps.gov/valr
www.pacifichistoricparks.org

Websites encompassing National Park Service World War II-related sites in Alaska, California, Hawaii, Guam, and Saipan. Information for USS *Arizona* Memorial tickets can be found here.

NavSource Naval History
www.navsource.org

An incredible amount of information and photos of every conceivable US naval vessel are hosted on this website. There are often special features centered around important anniversary dates in US Navy history.

North Valley Diver Publications
www.northvalleydiver.com

Dan E. Bailey's website for his outstanding books and detailed maps for divers interested in Truk Lagoon and Palau.

Ocean Riders Inc.
www.mauioceanriders.com/lanai.html

The only snorkel tour that circumnavigates the island of Lanai using thirty-foot-long, rigid-hull inflatable boats and one of the only ways to see Shipwreck Beach from the water.

Odyssey Marine Exploration
www.shipwreck.net

Worldwide leader in deep-ocean exploration and pioneer of deep-sea archaeological shipwreck excavations. Recovers of tons of gold and silver from a variety of shipwrecks. Be sure to visit their archive of archaeological papers.

On Eternal Patrol
www.oneternalpatrol.com

This site provides information on submarines lost and includes a directory of "Lost Submariner Family Networks." There are more than 4,100 memorial pages for those lost in the US submarine service.

Pacific Wrecks
www.pacificwrecks.com

A nonprofit organization devoted to sharing information about the Pacific theater of World War II and the Korean War. Includes message board and profiles of many Allied and Axis ships sunk during World War II.

Sharkhunters International
www.sharkhunters.com

The organization preserves many of the officers' and sailors' firsthand accounts of the submarine war of all nations during World War II. Sharkhunters has a heavy slant toward U-boats.

Brad Sheard (Photography by Brad Sheard)
www.bradsheard.com

Author and photographer Bradley Sheard's website presents many of his full-color wreck photos and articles about his diving adventures. There are also links to the books he's written.

Shipbuilding History
www.shipbuildinghistory.com

This website hosts the construction records of the United States and Canadian shipbuilders and boat builders during World War II.

The Sub Pen

www.thesubpen.com

Submarine and U-boat history website.

U-Boat.net

www.uboat.net

Everything there is to know about U-boats, the Allied ships attacked, technological information, submarine commanders, and the fates of the boats.

U-Boat Archive

www.uboatarchive.net

Excellent research site detailing every U-boat with extensive photos and scans of many original documents.

U-Boat Story

www.u-boatstory.co.uk

See *U-534* up close, then explore the interactive museum and view artifacts and personal effects recovered from inside the submarine.

US Naval History and Heritage Command

www.history.navy.mil

Excellent resource for US Navy information and photos including fleet information, museums, a digital copy of *Dictionary of American Naval Fighting Ships*, and much more.

US Naval Institute

www.usni.org

Publishers of *Proceedings* and *Naval History* magazines.

US Navy Sunken Military Craft Act

www.history.navy.mil/branches/org12-12a.htm#ftn1

Full text of H.R. 4200, the Ronald W. Reagan National Defense Authorization Act for Fiscal Year 2005, more commonly referred to as Title XIV–Sunken Military Craft. This is the law that governs all sunken US vessels. It states: "Right, title, and interest of the United States in and to any United States sunken military craft– (1) shall not be extinguished except by an express divestiture of title by the United States; and (2) shall not be extinguished by the passage of time, regardless of when the sunken military craft sank."

World War II Maritime Heritage Trail—Battle of Saipan

www.pacificmaritimeheritagetrail.com

Information on diving sunken Allied and Japanese shipwrecks, aircraft, and tanks around the island of Saipan. Outstanding photos accompany the interpretive information.

SHIP WEBSITES

Historic Naval Ships Association (HNSA)

www.hnsa.org

Extensive resources and Internet guide to preserved historic ships around the world. An excellent starting place for anyone interested in nautical history.

Aircraft Carrier

USS *Hornet* (CV-12) • www.hornet-museum.org

USS *Intrepid* (CV-11) • www.intrepidmuseum.org

USS *Lexington* (CV-16) • www.usslexington.com

USS *Midway* (CV-41) • www.midway.org

USS *Yorktown* (CV-10) • www.patriotspoint.org

Battleship

USS *Alabama* (BB-60) • www.ussalabama.com

USS *Iowa* (BB-61) • www.pacificbattleship.com

USS *Massachusetts* (BB-59) • www.battleshipcove.org

USS *Missouri* (BB-63) • www.ussmissouri.org

USS *New Jersey* (BB-62) • www.battleshipnewjersey.org

USS *North Carolina* (BB-55) • www.battleshipnc.com

USS *Texas* (BB-35) • www.tpwd.state.tx.us/state-parks/battleship-texas

USS *Wisconsin* (BB-64) • www.nauticus.org

Cruiser

USS *Little Rock* (CL-92) • www.buffalonavalpark.org

USS *Salem* (CA-139) • www.uss-salem.org

Destroyer/Destroyer Escort

USS *Cassin Young* (DD-793) • www.nps.gov/bost/historyculture/usscassinyoung.htm

USS *Joseph P. Kennedy, Jr.* (DD-850) • www.battleshipcove.org

USS *Kidd* (DD-661) • www.usskidd.com

USS *Laffey* (DD-724) • www.patriotspoint.org

USS *Orleck* (DD-886) • www.orleck.org

USS *Slater* (DE-766) • www.ussslater.org

USS *Stewart* (DE-238) • www.americanunderseawarfarecenter.com

USS *The Sullivans* (DD-537) • www.buffalonavalpark.org

Landing Craft

USS LCI(L)-102 • www.mightymidgets.org

USS LCI(L)-713 • www.amphibiousforces.org

USS LCI(L)-1091 • www.redwoods.info

USS LCS(L)(3)-102 • www.mareislandhpf.org/ships.html

USS LST-325 • www.lstmemorial.org

USS LST-393 • www.lst393.org

Liberty Ship

SS *Jeremiah O'Brien* • www.ssjeremiahobrien.org

SS *John W. Brown* • www.ssjohnwbrown.org

PT Boat (see table on page 206)

PT 658 • www.savetheptboatinc.com

Submarine

USS *Batfish* (SS-310) • www.ussbatfish.com

USS *Bowfin* (SS-287) • www.bowfin.org

USS *Cavala* (SS-244) • www.americanunderseawarfarecenter.com

USS *Clamagore* (SS-343) • www.patriotspoint.org

USS *Cobia* (SS-245) • www.wisconsinmaritime.org

USS *Cod* (SS-224) • www.usscod.org

USS *Drum* (SS-228) • www.drum228.org

USS *Ling* (SS-297) • www.njnm.org

USS *Lionfish* (SS-298) • www.battleshipcove.org

USS *Pampanito* (SS-383)* • www.maritime.org

USS *R-12* (SS-89) • www.r12sub.com

USS *Razorback* (SS-394) • www.aimmuseum.org/uss-razorback

USS *Requin* (SS-481) • www.carnegiesciencecenter.org

USS *Silversides* (SS-236) • www.silversidesmuseum.org

USS *Torsk* (SS-423) • www.usstorsk.org

Tug

YT-146 Hoga • aimmuseum.org/uss-hoga

US Coast Guard

USCGC *Ingham* (WHEC-35) • www.uscgcingham.org

USCGC *McLane* (WMEC-146) • www.silversidesmuseum.org

USCGC *Taney* (WHEC-37) • www.historicships.org/taney.html

Victory Ship

SS *American Victory* • www.americanvictory.org

SS *Lane Victory* • www.lanevictory.org

SS *Red Oak Victory* • www.ssredoakvictory.com

DIVING MAGAZINES WITH WORLD WAR II WRECK COVERAGE

Advanced Diver Magazine • www.advanceddivermagazine.com

Diver Magazine • www.divermag.com

Sea Classics Magazine • www.challengeweb.com/sea-classics.html

Scuba Diving • www.scubadiving.com

Sport Diver • www.sportdiver.com

Wreck Diving Magazine • www.wreckdivingmag.com

INDEX

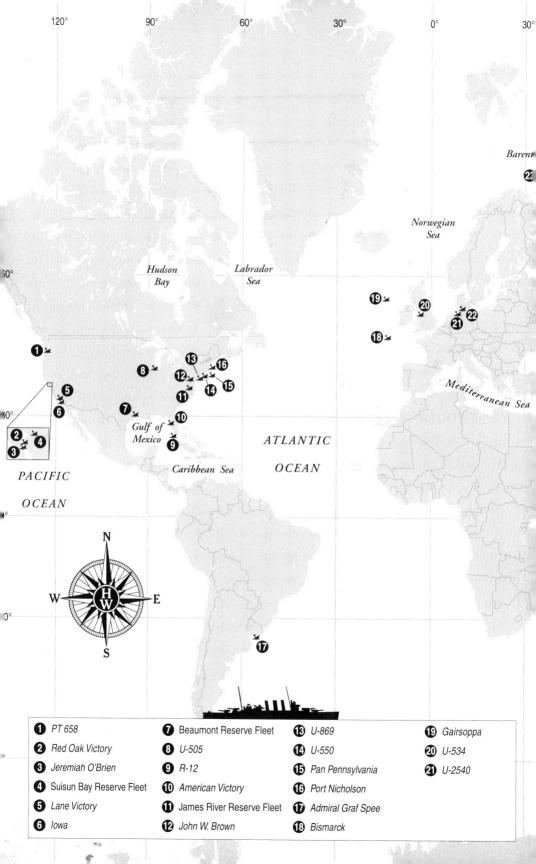

120°   90°   60°   30°   0°   30°

Baren[

Norwegian
Sea

Hudson
Bay

Labrador
Sea

60°

PACIFIC

OCEAN

Gulf of
Mexico

Caribbean Sea

ATLANTIC

OCEAN

Mediterranean Sea

N

W   H   E
    W

S

①	PT 658	⑦	Beaumont Reserve Fleet	⑬	U-869	⑲	Gairsoppa
②	Red Oak Victory	⑧	U-505	⑭	U-550	⑳	U-534
③	Jeremiah O'Brien	⑨	R-12	⑮	Pan Pennsylvania	㉑	U-2540
④	Suisun Bay Reserve Fleet	⑩	American Victory	⑯	Port Nicholson		
⑤	Lane Victory	⑪	James River Reserve Fleet	⑰	Admiral Graf Spee		
⑥	Iowa	⑫	John W. Brown	⑱	Bismarck		